FAKE INVISIBLE CATASTROPHES AND
THREATS OF DOOM

Also by Patrick Moore:

*Confessions of a Greenpeace Dropout -
The Making of a Sensible Environmentalist*

Trees are the Answer

Copyright © 2021 Patrick Moore
All rights reserved
Published by Ecosense Environmental Inc.
Printed by Amazon

No part of this book, covered by copyright, may be reproduced or used in any form or by any means (graphic, electric, or mechanical, including photocopies, taping or information storage and retrieval systems) without the prior written consent of the publisher.

Library and Archives Canada Cataloguing in Publication

Title: Fake invisible catastrophes and threats of doom / Patrick Moore.
Names: Moore, Patrick, 1947- author. | Ecosense Environmental Inc., publisher.
Description: Includes bibliographical references and index.
Identifiers: Canadiana 20200406035 | ISBN 9798568595502 (softcover)
Subjects: LCSH: Global environmental change. | LCSH: Climatic changes. | LCSH: Carbon dioxide—Environmental aspects. | LCSH: Nature—Effect of human beings on. | LCSH: Environmentalism.
Classification: LCC GE149 .M66 2021 | DDC 363.7—dc23

Includes bibliographical references
ISBN 979-85685955-0-2

1. Moore, Patrick Albert 1947 -. 2. Environmental Science. Climate Change. Nuclear Energy. Extinction. Genetic Science. Title.

For ordering information contact:

Ecosense Environmental
2080 Beach Drive, Comox, B.C., Canada V9M 1T8
Phone: (604) 220-6500
email: pmoore@ecosense.me
Internet: www.ecosense.me

Book Designed by CircularStudio.com

FAKE INVISIBLE CATASTROPHES AND THREATS OF DOOM

Patrick Moore

The Sensible Environmentalist

Contents

INTRODUCTION 7

CHAPTER I
Africa's Oldest Baobab Trees are Dying 15

CHAPTER II
The Great Barrier Reef is Dead or Dying and All Coral Reefs Will Die in this Century 19

CHAPTER III
Climate of Fear and Guilt 31

CHAPTER IV
Polar Bears are Threatened with Extinction Because of Climate Change 80

CHAPTER V
One Million Species Face Extinction Due to Climate Change – Soon 95

CHAPTER VI
The Great Pacific Garbage Patch is Full of Plastic and is Twice the Size of Texas 103

CHAPTER VII
Genetically Modified Foods Contain Something Harmful. What is it? *129*

CHAPTER VIII
Fear of Invisible Radiation from Nuclear Energy *145*

CHAPTER IX
Forest Fires: Of Course They are Caused by Climate Change (Not Trees?) *161*

CHAPTER X
Ocean Acidification – A Complete Fabrication *173*

CHAPTER XI
Mass Walrus Death from CO_2 – Another Fake Catastrophe from Sir David Attenborough *198*

EPILOGUE *206*

Introduction

As a lifelong learner I've found my time on Earth has been filled with a series of longish, mostly enjoyable interludes punctuated by the occasional flash of revelation. These are the Eureka moments, the ones that can shape one's life and change its direction and permanently alter the way one sees the world.

At the age of fifteen, my mother Beverly introduced me to the philosophical writings of Bertrand Russell. My first adult insight upon reading his book, *Authority and the Individual*, was that politics can be summed up as a contest between those seeking control over society and those seeking freedom from society's control. Both are necessary objectives, but it is gaining a healthy balance between the two that is most important. This, coupled with the fact that one person's idea of balance is another person's perception of bias, makes it all the more interesting, and volatile.

In 1965, during my second year of forestry at the University of British Columbia (UBC), I attended a noontime guest lecture by Dr. Vladimir Krajina – the former Deputy Minister of Forests in Czechoslovakia, until he was forced to flee from the Russian communists shortly after World War II. It was there that I heard the term "ecology" used for the first time. It was a word not used in the popular press, even though "the environment" was already a common topic. Since its development in the late 1800s, ecology had been an obscure branch of biology, its inception originating from the study of Ukrainian grasslands soils. With the advent of environmentalism, ecology was about to take control of a big piece of modern thought.

Ecology is essentially the study of how all the components of our world interrelate and interact together. Specifically, how living things, rocks, soils, water, and air interrelate and interact. This is pretty close to an infinity of interrelationships among the nearly infinite number of components. Until 1965, I'd been raised in a decidedly agnostic family in the woods of northern Vancouver Island where the concept of "spirituality" was not usually part of our dinner conversation. At the time, I thought science was a purely technical subject where everything could be measured and quantified. I suddenly came to learn that through the science of ecology, one could gain insight into the wondrous infinity of life, and of the universe. Upon completing my Bachelor of Science with Honors, I enrolled in a PhD program in Ecology and never looked back. I became a born-again ecologist.

In the late 1960s the world experienced various social revolutions; it was the height of the Cold War, the Vietnam War, as well as the threat of all-out nuclear annihilation. It was during this combination of anxiety-producing circumstances that the newly emerging consciousness of the environment blossomed. These were some of the most fear-inducing years of our era. I remember being motivated to actually do something about it, something beyond simply reading books about ecology and writing exams on the topic. In early 1971, in the midst of my PhD program at UBC, I joined a small group called the Don't Make a Wave Committee that met in the basement of the Unitarian Church in Vancouver to plan a protest voyage against US underground hydrogen-bomb testing in Alaska. I sailed as the lone ecologist on board, on a mission against what was arguably the most powerful organization on Earth at the time – the US Atomic Energy Commission. It was on this mission that, together, we became the organization known as Greenpeace. And it was Greenpeace that consumed my next 15 years. It was quite a journey, all of which is covered in my previous book, *Confessions of a Greenpeace Dropout – The Making of a Sensible Environmentalist*, along with a chapter in this book on each of the environmental issues of concern today.[1]

---

1. Patrick Moore, *Confessions of a Greenpeace Dropout – The Making of a Sensible Environmentalist*, Beatty Street Publishing, 2013, pp454. https://www.amazon.com/Confessions-Greenpeace-Dropout-Sensible-Environmentalist-ebook/dp/B004X2I6ZM.

The next revelation in my life came to me in 1982 at a meeting of 85 international environmental leaders. Representatives were chosen from each continent, and all of us convened at the Environment Liaison Centre in Nairobi, Kenya. It was there that I heard about the recently coined concept "sustainable development." In more recent times, the term has been rendered practically useless by extremists on both sides of the argument. The side on the far right tends to think it is code for globalism and the far left side thinks it's a dangerous compromise. But at the time I first heard the term, sustainable development, it made a major impression on some of us in the growing environmental movement. Until then, we had never paid much attention to economic and social realities as we were so entirely focused on the environmental agenda. Many still don't adequately consider social and economic realities in their environmental policies. The definition of sustainable development is relatively straightforward; it is defined as "an effort to find the appropriate balance among environmental, social, and economic priorities," however, not necessarily in that order. To many current environmental activists, "the people" are a kind of afterthought, to be considered only once the perfect environmental policy has been identified, without regard for how negative the impact is on humanity.

Sustainable development means we have to consider the social and economic priorities of the people if we want to curb negative environmental activities. This consideration parallels the social revolution from more than 100 years earlier when child labor was outlawed and when women first received the right to vote. These social transitions had to be incorporated into the economic structure of that time. During the past 50 years we have adopted a lot of environmental policies that have changed the social and economic landscape considerably. But today there are demands being made that would actually cripple society and the global economy permanently. The push to "phase out all fossil fuel consumption in thirty years" is certainly the biggest threat to civilization in the world today.

In the mid 1980s, I finally decided to leave Greenpeace due to their transition from what was sensible environmentalism, to a platform of anti-human and anti-science campaigns that were more concerned with fundraising and scaring people with misinformation than with improving the environment. The adoption of the campaign to ban chlorine (the devil's element) worldwide in 1986 by my fellow directors of Greenpeace International, none of whom had any formal science education, was the final straw and for me signaled my departure. The grounds for my parting ways were based on the fact that chlorine is the most important of all the 94 natural elements for both public health and medicine. Adding chlorine to drinking water, swimming pools, and spas was the biggest advance in the history of public health.[2] And moreover, more than 85 percent of our prescription medications are made using chlorine chemistry.[3] Twenty-five percent of our medications actually contain chlorine. And then there's polyvinyl chloride, also known as PVC or simply vinyl, the most versatile of all the plastics.[4]

It was bittersweet parting ways with Greenpeace, the organization that I had helped build, shape, and guide for 15 years. Unfortunately, Greenpeace had gone from an altruistic group of volunteers with a noble vision, to a business with an ever-expanding budget, a matching payroll to meet, and was now rapidly transforming into a racket peddling junk science.

During my last years with Greenpeace I had taken an interest in the newly emerging science of marine aquaculture, and in 1984 I was given a book: *Seafarm – The Story of Aquaculture*. It was that publication that gave me the next path I was looking for.[5] My childhood home of Winter Harbour on northern Vancouver Island offered the right habitat for salmon farming which had already taken off in much of Norway. Aquaculture was now in its infancy on the coast of British Columbia. My brother, Michael, and brother-in-law, Peter, joined my wife Eileen and I in a family business. We built a fish hatchery and spent the next eight years growing Chinook (king) salmon in the sea.

It is now clear that marine and freshwater aquaculture will become one of the most important industries providing healthy oils and proteins to people worldwide. When we began salmon farming in 1985, aquaculture provided for about 10 percent of the world's seafood consumption. Today it is fast approaching 50 percent and will continue to climb as offshore aquaculture is developed.[6]

I suppose it was inevitable that I would eventually transition back into the global environmental discussion. This time, however, in the role of helping to guide governments and industries into sensible policies that would improve their environmental performance without

---

2. Daryl Weatherup, "Chlorine: Protecting human health for more than a century," *GreenBiz*, November 14, 2016. https://www.greenbiz.com/article/chlorine-protecting-human-health-more-century.
3. "Chlorine Chemistry – Providing Pharmaceuticals that Keep You and Your Family Healthy," American Chemistry Council, 2020. https://chlorine.americanchemistry.com/Chlorine/Medicine/.
4. "An Introduction to Vinyl," AZO Materials, October 25, 2001. https://www.azom.com/article.aspx?ArticleID=987.
5. Elisabeth Mann Borgese, *Seafarm: The story of aquaculture*, Harry N. Abrams January 1, 1980. https://www.amazon.com/Seafarm-aquaculture-Elisabeth-Mann-Borgese/dp/0810916045.
6. United Nations Food and Agriculture Organization, *The State of World Fisheries and Aquaculture 2018*, http://www.fao.org/3/i9540en/i9540en.pdf.

driving them into bankruptcy. Much of this practice involved learning to differentiate between demands that are based on good science and knowledge as opposed to those that are primarily aimed at merely fundraising or damaging the economy with dubious or even malevolent intentions.

With my background now laid out, I move on to the central thesis of this book:

Awhile back it dawned on me that the great majority of scare stories about the present and future state of the planet, and humanity as a whole, are based on subjects that are either invisible, like CO_2 and radiation, or extremely remote, like polar bears and coral reefs. Thus, the vast majority of people have no way of observing and verifying for themselves the truth of these claims predicting these alleged catastrophes and devastating threats. Instead, they must rely on the activists, the media, the politicians, and the scientists – all of whom have a very large financial and/or political stake in the subject – to tell them the truth. This welcomes the opportunity to simply invent narratives such as the claim that "CO_2 emissions from burning fossil fuels are causing a climate emergency."

No one can actually see, or in any way sense, what CO_2 might actually be doing because it is invisible, odorless, tasteless, silent, and cannot be felt by the sense of touch. Therefore, it is difficult to refute such claims because there is nothing to point to and tangibly expose the falsity of these claims. One can't simply point to visible CO_2 and say, "Look what awful things CO_2 is doing over there." Thus, CO_2 as a harmful, world-ending emission, is an almost perfect subject to invent and propagate a doomsday story, and this fact has not gone unnoticed by those inclined to peddle unsubstantiated fabrications. CO_2 has become the scapegoat of blame for an entire laundry list of negative effects, that could require 118 books to record and tabulate. And indeed, the website goodreads.com lists 118 books on the subject of climate change; and that's confined to books exclusively written in the English language.[7]

When one studies these "narratives" of invisible and remote circumstances, it is hard to avoid noticing that the purveyors often stoop to ridiculing and shaming, and likewise exhibit an unwillingness to discuss the allegations in a civilized manner. It is virtually impossible to engage in debate as they usually dismiss those who question their narrative as a skeptic, liar, denier, or of being in the pockets of "big oil." And if the alleged skeptic has employment, these narrators will work

---

7. https://www.goodreads.com/list/show/39667.Best_Climate_Change_Books.

underhandedly to have you removed from your livelihood or position. In summary, these purveyors of global environmental catastrophes are definitely a scurrilous and dishonest lot. Healthy skepticism is at the very heart of scientific inquiry, and it has played an integral role in determining factual scientific truth. It is the duty of scientists to be skeptical of all new claims, especially when they are predictions of catastrophes that have not yet occurred.

The scientific method is not that complicated. In most religions we are asked to "believe" even though we have not directly observed the alleged higher powers or beings. However, in science, direct and tangible observation is fundamental. We must "observe" the situation with our senses of sight, smell, taste, hearing, and/or touch. Or, observations can be made using an instrument such as the microscope, telescope, Geiger counter, voltmeter, X-ray image, etc. Upon observing a new phenomenon, we must repeat the observation under the same conditions numerous times in order to "verify" that it is a repeatable occurrence and not just a fluke. It is only then that we should make an announcement of our hypothesis and challenge other scientists to "replicate" and test our results. If enough replications under similar circumstances are demonstrated by other scientists, replications that produce the same result, we are verging on a theory in science. Observation, verification, replication.

I am purposefully placing the chapter on climate change part-way through the book, so that some of the unavoidable technical details that climate change involves do not immediately cause fatigue. Many of the following chapters deal with claims made about the effects of climate change, such as the impacts on polar bears, coral reefs, species extinction, trees and forests, etc. Other chapters consider claims made that are not directly related to CO_2 such as genetic modification, chemicals in the environment, plastics, and radiation (nuclear energy). Let's get started with a classic example that is actually humorous because it is so ridiculous.

CHAPTER ONE

Africa's Oldest Baobab Trees are Dying

A **headline** in *USA Today* reads "Africa's Oldest Baobab Trees are Dying at an Unprecedented Rate, and Climate Change may be to Blame."[1] Just in case you think I'm kidding here is the masthead:

NEWS

Africa's oldest baobab trees are dying at an unprecedented rate, and climate change may be to blame

Doyle Rice USA TODAY
Published 2:12 p.m. ET Jun. 11, 2018 | Updated 2:34 p.m. ET Jun. 11, 2018

What if you read a headline stating, "China's oldest people are dying"? Wouldn't that be perfectly normal, and shouldn't the same go for every species including trees? But then there is the claim of

---

1. Doyle Rice, "Africa's Oldest Baobab Trees are Dying at an Unprecedented Rate, and Climate Change may be to Blame," *USA Today*, June 11, 2018. https://www.usatoday.com/story/news/2018/06/11/baobab-trees-dying-africa-climate-change/690946002/.

Figure 1. The Avenue of the Baobabs in Madagascar helps illustrate these trees' unique form and why they are actually nick-named the "upside down trees," as they appear to have been uprooted and stuck back into the ground with their roots pointing upwards.

"unprecedented rate" which implies that baobab trees may be on the way out if they are dying faster than they are being born. No mention is made of the rate that they were dying before it became unprecedented. That in itself is enough to suspect fakery. Notice the word "may" appears in the subhead. In order to be truthful it should say "may or may not" as "may" implies conjecture rather than being a statement of fact.

Baobab trees are certainly not invisible, but no one is blaming the baobabs for their rate of dying. It is that evil CO_2 causing the climate in Africa to change that is being blamed. No evidence for this is given in the story that actually links climate change to dead baobabs; it's probably just a hunch.

In my own search, I could not find a single reference for how many baobab trees there are growing in Africa; but they have a huge range, across the continent below the Sahara Desert and down the east coast to South Africa (see Fig. 2). How is it possible to determine the "rate" at which the trees are dying if there is no inventory of their population? Baobabs are also native to India, Madagascar, and a number of Indian Ocean islands.

Regarding the "unprecedented" claim there was only one alleged fact provided. Adrian Patrut, a chemist at Babes-Bolyai University in Romania, headed the study of 60 older Baobab trees which were carbon-dated. The oldest was said to be 2,500 years old. For comparison

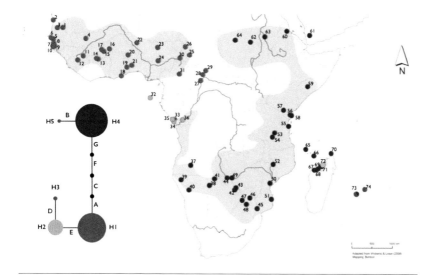

Figure 2. The natural range of the baobab tree in Africa; an area nearly as large as the lower 48 states of the United States. There are at least tens of thousands and perhaps hundreds of thousands of baobab trees on the continent of Africa.

the giant sequoia in California can live for up to 3,500 years. I imagine the oldest trees in their population are dying too.

The data regarding the claim of an unprecedented number of trees dying was that "eight of the thirteen oldest trees have died in the past thirteen years."[2] That is just more than one tree dying every two years. If there were only 10,000 baobab trees and they lived an average of 1,500 years, then 6.7 trees could die every year and the population would remain stable. It is understandable, due to their unique form and grand size, that people would be concerned for the well-being of baobab trees. But the story is so obviously fake that it does not take a genius to see through it. Despite this, even a Fox affiliate ran with this story,[3] along with close to 150 other news outlets around the world.

---

2. Rachel Nuwer, "Africa's Ancient Baobabs Are Dying, Researchers fear the trees are parched by drought and rising temperatures linked to climate change," the *New York Times*, June 12, 2018. https://www.nytimes.com/2018/06/12/science/baobabs-climate-change-drought.html?auth=login-google.
3. "Ancient Baobab trees in Southern Africa are dying: Scientists blame climate change," *Fox Six Now*, December 32, 2018. https://fox6now.com/2018/12/31/ancient-baobab-trees-in-southern-africa-are-dying-scientists-blame-climate-change/.

Figure 3. Many of the publications that ran with the "baobabs are dying" story included a picture similar to the one above. Most people think of tropical trees as evergreen whereas the baobab is deciduous and loses all its leaves in October. This can give the appearance that they are dead or dying, especially when placed in a story claiming the baobabs are dying. Very few of the articles I found through researching this issue contained a photo of an actual dead baobab tree, and none considered the possibility that old age might be the cause.

CHAPTER TWO

The Great Barrier Reef is Dead or Dying and All Coral Reefs Will Die in this Century

Coral reefs, the Great Barrier Reef in particular, are a good example of a supposed ongoing catastrophe that is blamed on invisible CO_2 emissions. These reefs are under the ocean's surface, invisible to nearly everyone, and most of them are in remote offshore locations in the tropics. Coral reefs are a global icon that everyone loves even if they have only seen them on TV, and they are the perfect subject for a fabricated catastrophe.

❚HUFFPOST❚

ENVIRONMENT 04/21/2016 06:14 pm ET

93 Percent Of The Great Barrier Reef Is Practically Dead

Climate change is destroying Earth's largest coral ecosystem.

By Chris D'Angelo

Figure 4. Many of the alarmist headlines about the Great Barrier Reef after the bleaching event in April 2016, cleverly used terms such as "dying," "practically dead," "bleached," and "terminal." None of which actually mean "dead." Carbon dioxide and climate change were universally blamed for this "catastrophe."

In April of 2016 there were headlines around the world claiming that 93 percent of the Great Barrier Reef (GBR) was dead, dying, bleached, or about to die. The *Huffington Post* headline read, "93 Percent of the Great Barrier Reef is Practically Dead: Climate change is destroying Earth's largest coral ecosystem."[1] What precisely does "practically dead" mean? Clearly it doesn't mean "dead," but it does the trick. Similar funeral headlines appeared in hundreds of news outlets around the world.[2] "But the careful researcher would be hard pressed to find the origin of the number 93 percent as it appeared only once in the text of a media release from the ARC Centre[3]. It claimed that "93% of the reef was 'affected'" (as opposed to bleached or dying) and then provided a graph that did not support this claim. Here is the graphic that accompanied the story in the *Huffington Post* (see Fig. 5).

As you can see from Figure 5, the GBR has been divided into three sections: northern, central, and southern. There are 911 individual reefs among them. There are only two categories of data provided for bleaching: severely bleached and not bleached. There is also no indication of how severe "severely" is. Is it 25 percent or 50 percent? Or 90 percent? Whatever it means there is no arithmetical way of arriving at the number "93 percent of the GBR" based on the figures from this graphic.

The source for this story was the ARC Centre of Excellence for Coral Reef Studies and the spokesperson quoted was Terry Hughes, convenor of the National Coral Bleaching Taskforce. One might expect that such authorities might know that the bleaching of corals is not synonymous with the death of corals, and that perhaps an astute environmental reporter might know this too. Yet, the distinct impression given was that the corals were dead or dying, and the immediate cause was bleaching.

Most corals are a symbiotic relationship between an animal and a phytoplankton, which is a tiny aquatic photosynthetic plant. The animal is called a polyp and is a relative of jellyfish and sea anemones. Most corals contain hundreds to hundreds of thousands of polyps. The plankton is referred to as a symbiont (also called a zooxanthellae).

---

1. Chris D'Angelo, "93 percent of the Great Barrier Reef is Practically Dead. Climate change is destroying Earth's largest coral ecosystem," *Huffington Post*, April 21, 2016. https://www.huffpost.com/entry/climate-change-destroying-great-barrier-reef_n_571918e6e4b0d912d5fde8d4.
2. Chris Mooney. "'And then we wept': Scientists say 93 percent of the Great Barrier Reef now bleached," *Washington Post*, April 20, 2016. https://www.washingtonpost.com/news/energy-environment/wp/2016/04/20/and-then-we-wept-scientists-say-93-percent-of-the-great-barrier-reef-now-bleached/
3. ARC Centre of Excellence - Coral Reef Studies, "Only 7% of the Great Barrier Reef has avoided coral bleaching." April 20, 2016. https://www.coralcoe.org.au/media-releases/only-7-of-the-great-barrier-reef-has-avoided-coral-bleaching.

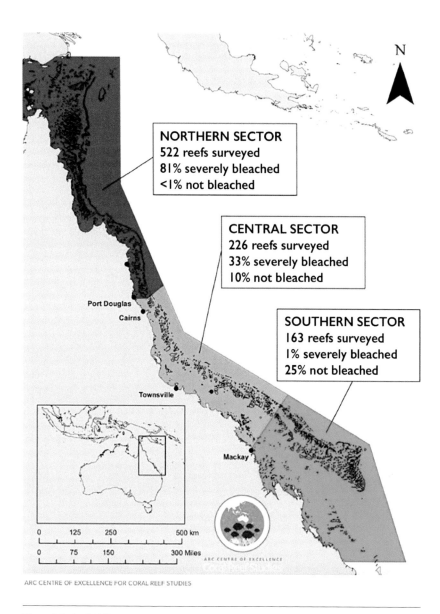

Figure 5. In April 2016, many media reports of the Great Barrier Reef used "93 percent" as the amount that the reef was dead, dying, nearly dead, etc. It is not actually known where this percentage came from.

Each polyp selectively ingests millions of the symbiont plankton into its tissue thus providing the plankton with protection from grazers. The plankton in turn produces sugars from carbon dioxide and water, some of which are given to the polyp as food in return for the shelter.

The coral polyps are colorless and transparent, much like some jellyfish, and it is the plankton that give the coral its colors including shades of green, red, brown, blue, and yellow, depending on the species of plankton ingested. It is quite common that in periods of warmer or colder seawater the polyps eject the plankton and therefore the coral colony becomes white because the polyps are transparent and the coral structure which is made of calcium carbonate (limestone) is white (see Fig. 6). The term "bleaching" has been adopted due to the whiteness of the coral but has nothing to do with actual bleaching, as would occur from sunlight or from chemicals such as sodium hypochlorite. Most people are familiar with the common knowledge of bleach used as a household item for whitening fabrics, and that it is also a disinfectant that kills bacteria. They are therefore more inclined to accept the idea that coral bleaching means that corals are dead or dying. That is, unless the explicit truth about what coral bleaching is and means is actually communicated by the scientists and the media. However, in this case neither the media nor the scientists made much effort to disclose this information. Corals very often survive bleaching events, and this is recognized by the US National Oceanic and Atmospheric Administration:

> *When water is too warm, corals will expel the algae (zooxanthellae) living in their tissues causing the coral to turn completely white. This is called coral bleaching. When a coral bleaches, it is not dead. Corals can survive a bleaching event, but they are under more stress and are subject to mortality.*[4]

Figure 6. An example of bleached coral. The polyps, which are transparent, are still alive and chances are good they will replace the symbiont phytoplankton when ocean conditions are to their liking. They may even spawn while they are bleached and therefore replace themselves in the event they do die.*

* "Bleached coral reefs showing new signs of life," *Daily Mercury*, October 6, 2017. https://www.dailymercury.com.au/news/bleached-coral-reefs-showing-new-signs-of-life/3231508/

---

4. NOAA, "What is coral bleaching?" National Ocean Service, NOAA, January 7, 2020. https://oceanservice.noaa.gov/facts/coral_bleach.html.

One of the most fascinating features of coral is that hundreds or even hundreds of thousands of polyps in the same coral colony appear able to communicate with each other as they all choose the same species of phytoplankton to ingest. It is quite common to see two colonies of the same species of coral living close by each other, each a different color due to having selected to ingest different-colored plankton species. It is hypothesized that the choice of plankton species is an adaptive behavior that allows the corals to adjust to warmer or cooler ocean temperatures.[5]

The next spate of news on the Great Barrier Reef came in April of 2017. This time the reef was declared "terminal" which like "practically dead" and "dying" doesn't mean "dead." But it surely indicates that death is just around the corner.[6,7]

Forbes' headline used the phrase "Final 'Terminal Stage,'" implying there were other terminal stages before this final one, but that this terminal stage would definitely lead to death (see Fig. 7). So, there you go, the Great Barrier Reef is dead and gone. Well not exactly.

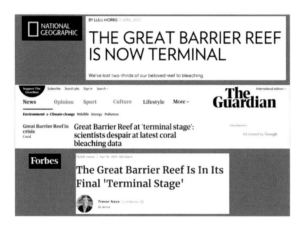

Figure 7. The 2017 coverage of the Great Barrier Reef's imminent demise was not as widespread as in 2016, but it did have heavier ominous overtones. Millions of people believed these reports even though they were purposely fabricated to support the "climate catastrophe" narrative, which itself has no basis in genuine science.

---

5. Jim Steele, "The Coral Bleaching Debate: Is Bleaching the Legacy of a Marvelous Adaptation Mechanism or A Prelude to Extirpation?" May 17, 2016. http://www.landscapesandcycles.net/coral-bleaching-debate.html.
6. Trevor Nace, "The Great Barrier Reef is in Its Final 'Terminal Stage,'" *Forbes*, April 15, 2017. https://www.forbes.com/sites/trevornace/2017/04/15/the-great-barrier-reef-is-its-final-terminal-stage/#3a21ff6f3f2e.
7. Christopher Knaus and Nick Evershed, "Great Barrier Reef at 'terminal stage': scientists despair at latest coral bleaching data," the *Guardian*, April 9, 2017. https://www.theguardian.com/environment/2017/apr/10/great-barrier-reef-terminal-stage-australia-scientists-despair-latest-coral-bleaching-data.

Eighteen months after the certain terminal demise of the reef was confirmed worldwide, some surprising headlines appeared. They were not as abundant as the previous year's obituaries, but they were still very explicit. On September 9, 2018, *Bloomberg* headlined: "Great Barrier Reef Showing 'Signs of Recovery': After a mass coral bleaching in 2016, the world's largest living structure is showing signs of a comeback."[8] And shortly after *Newsweek* reported that the "Great Barrier Reef Definitely Not Dead: Experts Announce Significant Signs of Recovery After Mass Bleaching."[9] Miracles apparently do happen (see Fig. 8). But the news of the recovery was nowhere near as widespread as the reports of the reef's demise. To this day I believe the majority of the public are quite certain the reef is dead or dying.

It seems many people tend to forget that every living creature eventually dies, and that more often than not, others are born to take their place. It certainly is newsworthy when there is a mass death, whether of people or of corals, but there is a big difference between mass death and extinction. The former allows for recovery, the latter not so much.

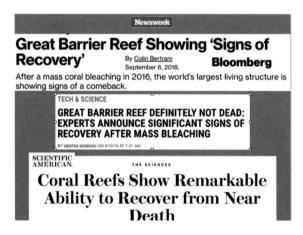

Figure 8. Very few media outlets ran this story, which means very few people were informed about the reef's recovery. The extent of the damage was clearly exaggerated from the start.*

* Joanne Nova, "Great Barrier Reef scare: exaggerated threats says head of GBR Authority," JoNova, June 6, 2016. http://joannenova.com.au/2016/06/great-barrier-reef-scare-exaggerated-threats-says-head-of-gbr-authority/.

---

8. Colin Bertram, "Climate Changed: Great Barrier Reef Showing 'Sign of Recovery.' After a mass coral bleaching in 2016, the world's largest living structure is showing signs of a comeback," *Bloomberg*, September 6, 2018. https://www.bloomberg.com/news/articles/2018-09-06/great-barrier-reef-showing-signs-of-recovery.
9. Aristos Georgiou, "Great Barrier Reef Definitely Not Dead: Experts Announce Significant Signs of Recovery After Mass Bleaching," *Newsweek*, September 10, 2018.https://www.newsweek.com/great-barrier-reef-definitely-not-dead-experts-announce-significant-signs-1113415.

My wife Eileen and I are fortunate enough to have a home in Cabo Pulmo, Baja California Sur, Mexico. This tiny village on the Sea of Cortez is graced with the largest coral reef on the west coast of the Americas. In 1996, it was designated a Federal Marine Park and fishing was banned inside the boundaries. When we first visited there in 1999, we snorkeled on the reef and immediately decided to buy a lot back behind the beach and build a home there. We have snorkeled there on countless occasions throughout the past 20 years.

Unfortunately, only two years after we built there, a hurricane brought about 20 inches of rain in two days and this filled Pulmo Bay with fresh water to a depth of about 15 feet, killing all the coral within that depth. During the past 18 years the coral has recovered remarkably well and is now back to more than 50 percent of its original density. The deeper corals that survived the deluge have recolonized the entire area and due to the fishing ban, the fish biomass has more than tripled. It is a reminder that even after complete elimination of corals over a large area, they recover relatively quickly. This is true of virtually all Earth's ecosystems. If a forest is destroyed by fire, insects, or felled by loggers for timber, and then simply left to its own devices it will recover fully in due course.

Let's dig a little deeper into the contention that many coral reefs are doomed to die by mid-century, and all of them by 2100, due to climate change warming the oceans.[10] We begin with the evolution of corals, the ancestors of which emerged 535 million years ago when the oceans were much warmer than today. Corals are in the taxonomic phylum Cnidaria and are one of many classes of marine species that are able to crystalize calcium carbonate to make protective shells for themselves.[11] They have survived three ice ages, including the present Pleistocene, and many other cataclysms far more extreme than anything happening in the present era. The modern corals evolved from their ancestors 225 million years ago when the climate was far warmer than it is today, and it remained warmer until 2.6 million years ago when the Pleistocene Ice Age set in.

So, you might wonder, if modern corals evolved and survived for 225 million years when the climate was considerably warmer than it is today,

---

10. UNESCO, "Assessment: World Heritage coral reefs likely to disappear by 2100 unless CO_2 emissions drastically reduce," United Nations Educational, Scientific and Cultural Organization, June 23, 2017. http://whc.unesco.org/en/news/1676/.
11. "Marine biogenic calcification," Wikipedia, December 10, 2019. https://en.wikipedia.org/wiki/Marine_biogenic_calcification.

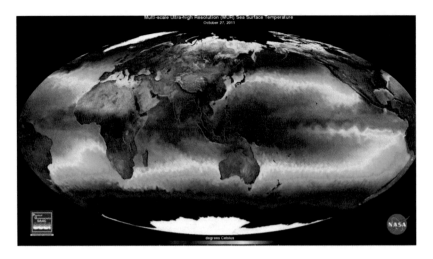

Figure 9. The Coral Triangle is clearly among the world's warmest ocean regions, if not the warmest. In addition, it is not subjected to seasonal colder waters from higher latitudes. It is home to more than 600 species of corals, including many soft corals found nowhere else, and to the highest diversity of reef fishes. Corals are limited in their range by colder ocean temperatures, not by warmer temperatures. The seas in the Coral Triangle are warmer than those at the Great Barrier Reef.*
* DAAC NASA, "The Multi-Scale Ultra-High Resolution (MUR) Sea Surface Temperature (SST) Data Set Animation," October 19, 2012. https://www.youtube.com/watch?v=1DNHRLgjLjA

why are we told that a small amount of warming threatens their very existence? Unfortunately, the answer is: for academic status and money.

The truth, however, is very clear. Among the warmest seas in the world are those in the Indonesian Archipelago, including the Philippines and the Solomon Islands. This is widely known as the Coral Triangle and is famous for having by far the world's highest biodiversity of coral, with more than 600 species, which is 76 percent of all coral species. The Coral Triangle also has the highest biodiversity of reef fish, with 2,000 species, which is 37 percent of all reef-fish species. In addition, the Coral Triangle is home to six of the world's seven species of marine turtles (see Fig. 9).[12]

The Coral Triangle therefore represents a kind of sanctuary for many of the coral and reef-fish species that would have occupied a much wider range when the Earth's climate was warmer than it is today. In other words, the reason the Great Barrier Reef, as well as other reefs, have fewer species of corals and reef fish than the Coral Triangle is because the oceans they occupy are not as warm as the oceans in the Coral Triangle.

---

12. "About the Coral Triangle," World Wide Fund for Nature, 2020. https://wwf.panda.org/knowledge_hub/where_we_work/coraltriangle/coraltrianglefacts/.

This interpretation is confirmed by a paper on global marine species diversity and the factors that influence higher or lower diversity. The paper concludes:

> In accordance with the idea of high kinetic energy (heat) facilitating greater species richness over evolutionary and ecological timescales, temperature emerged as the primary environmental correlate of diversity at the large geographic scale tested here.[13]

In other words, they found no evidence that there is anywhere in the world's oceans that indicate a decline in species richness due to warmer ocean water. They found the opposite, that the warmest waters in the world have the highest species diversity for every taxonomic class of marine life (see Figs. 10, 11, and 12, page 28). It is interesting to note this is also the case for terrestrial species: there is far more species diversity in the hottest tropical rainforest than there is in the Arctic or Antarctica. The authors go on to further conclude:

> Based on these findings, changes in the temperature of the global ocean may have strong consequences for the distribution of marine biodiversity.

And then, even though the authors found the highest biodiversity of virtually all marine species from 13 taxonomic groups in the warmest ocean environments, they surprisingly conclude that:

> Limiting the extent of ocean warming...may be of particular importance to securing marine biodiversity in the future.

It is simply not conceivable or logical to reach this conclusion from the facts presented in the study. If they had found that the warmest oceans had lower biodiversity than some that were cooler one could understand their musings. It appears the authors may simply wish to stay in the camp of those who project fear of warming.

The Scandalous Saga of Dr. Peter Ridd

The claims made by many scientists, mainly at universities, that coral reefs and especially the Great Barrier Reef are dying fast and may soon

---

13. Derek P. Tittensor, et al., "Global patterns and predictors of marine biodiversity across taxa," *Nature*, 466, August 26, 2010, pp1099-1101. https://www.nature.com/articles/nature09329.epdf

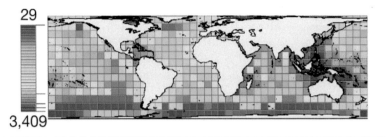

Figure 10. A map of biological diversity of all 13 taxa included in the study. The dark blue at 29 species to the dark red of 3,409 coastal marine species, including corals and reef fishes, clearly shows the region including Indonesia has the highest diversity, highly correlated with the warmest ocean water. This fact is ignored by climate alarmists despite the fact that the World Wide Fund for Nature, a leading climate-alarmist organization, admits it to be true.*

* Derek P. Tittensor, et al., "Global patterns and predictors of marine biodiversity across taxa," *Nature*, 466, August 26, 2010, pp1099-1101. https://wwf.panda.org/discover/knowledge_hub/where_we_work/coraltriangle/coraltrianglefacts/

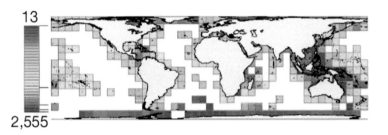

Figure 11. The relative biodiversity from 13 species to 2,555 species of coastal fishes is also highest in the Coral Triangle region centered on Indonesia.* Our Earth is in a relatively cold climate today compared to the past 500 million years. This has resulted in many species being restricted to a much smaller area than when the climate was considerably warmer in the past.
* Ibid.

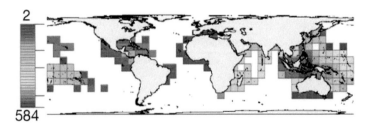

Figure 12. The distribution areas with only two species of corals to those with 584 species of corals.* Coral distribution and relative biodiversity indicate very clearly how important the Coral Triangle is for the survival of many coral species. Coastal corals are either very sparse or absent altogether along most of the world's coastlines.
* Ibid.

be extinct, has culminated in an opposing view from Dr. Peter Ridd, a 30-year faculty member of James Cook University in the state of Queensland, Australia. He contends that there is extreme exaggeration and misinformation in the claims of dying reefs and has publicly criticized his fellow scientists. Rather than encouraging a civil debate of the science, the university fired Dr. Ridd for not being "collegial"; in other words, he was fired for disagreeing with other scientists, which is supposed to be part of the scientific process called academic freedom. Here is what he had said in an article in the journal for the Institute of Public Affairs:

> *The basic problem is that we can no longer trust the scientific organizations like the Australian Institute of Marine Science, even things like the Australian Research Council Centre of Excellence for Coral Reef Studies…the science is coming out not properly checked, tested or replicated and this is a great shame because we really need to be able to trust our scientific institutions and the fact is I do not think we can any more.*[14]

Before firing him, the university issued Dr. Ridd with a gag order and told him he could not mention the case or the charges against him to anyone – not even to his wife. When the university found out he was receiving support from the Institute of Public Affairs, they searched his emails, found he had told his wife and colleagues about the case, charged him with 25 more violations, and then fired him. Dr. Ridd filed a lawsuit for wrongful dismissal and raised more than A$250 thousand on GoFundMe to fight his case.[15] The university spent more than A$600 thousand of taxpayer's money defending themselves for stripping Peter Ridd of his academic freedom and firing him. In September 2019, the Federal Circuit Court found the university guilty and awarded Dr. Ridd A$1.2 million, citing 18 violations of the law by James Cook University.

You might think that would end the matter, but not so. James Cook University appealed the decision to the Federal Court of Australia. This required Dr. Ridd to raise an additional A$600 thousand. The university spent approximately A$4 million on the appeal. Both the Premier of Queensland and the Prime Minister of Australia, who fund the

---

14. Peter Ridd, "Science or silence? My battle to question doomsayers about the Great Barrier Reef," Fox News, February 8, 2018. https://www.foxnews.com/opinion/science-or-silence-my-battle-to-question-doomsayers-about-the-great-barrier-reef.
15. GoFundMe, "Peter Ridd Legal Action Fund," Peter Ridd, July 29, 2020. https://ca.gofundme.com/f/peter-ridd-legal-action-fund-2019.

university and could have ended this with the stroke of a pen, remained silent while this travesty proceeded.

On July 22, 2020, the Federal Court of Australia overturned the Federal Circuit Court's decision on the judgement that the university's "Code of Conduct" overrode the guarantee of freedom of speech ensured in the university's "Enterprise Agreement." Dr. Ridd's "misconduct" was thus defined by his refusal to be gagged while being threatened by university officials, and by his public disagreement with other scientists who were misrepresenting the state of the Great Barrier Reef in order to bolster their massive research grants.

Presently, Dr. Ridd has completed the process of applying to the High Court, to appeal the decision of the Federal Court. This decision will be final for his case and will have a real impact on the standing of academic freedom and free speech in Australia, and possibly elsewhere. A decision is expected before Christmas 2020 and if accepted, the High Court will hear arguments sometime in 2021.[16] Fingers should be crossed.

---

16. Ibid.

CHAPTER 3
Climate of Fear and Guilt

Preamble

You have heard the news on climate change that says human-caused emissions of carbon dioxide are going to make the world too hot for life. So now as you drive down the highway in your SUV, you are afraid that you are killing your grandchildren by doing so. As this makes you feel guilty and accountable, you vow to send a hefty donation to Greenpeace, or any of the other hundreds of "charities," selling you this narrative. It is a very effective strategy on their part, as stirring a combination of fear and guilt is the most powerful motivator to get people to open their wallets in an effort to help avoid this alleged disaster. And all this inevitable doom due to an invisible gas that is essential for life and even now is only 0.0415 percent of the atmosphere.

Climate change is the most technically complicated subject among the many alleged catastrophes we are warned about regularly by alarmists. The climate of planet Earth has never stopped changing since the Earth's genesis, sometimes relatively rapidly, sometimes very slowly, but always surely. Hoping for a "perfect stable climate" is as futile as hoping the weather will be the same and pleasant, every day of the year, forever. So obviously we must accept that some climate change is always inevitable, and natural, whether it is through rising or falling temperatures, increased flooding or droughts, or more or fewer hurricanes and tornadoes.

Here is a list of just some of the predicted consequences as a result of climate change that have been drilled into anyone that dares turn on a TV, radio, or read a newspaper or magazine. Climate change caused by human emissions of CO_2 will cause:

Higher temperatures[1]
Lower temperatures[2]
More snow and blizzards[3]
Drought, fire, and floods[4]
Rapidly rising sea levels[5]
Disappearing glaciers[6]
Total loss of sea ice at the North Pole[7]
Mass species extinction[8]
More and stronger storms causing more damage[9]
Dying forests[10]
Death of coral reefs and shellfish[11]

---

1. Alan Buis, "A Degree of Concern: Why Global Temperatures Matter," NASA, September 23, 2020. https://climate.nasa.gov/news/2865/a-degree-of-concern-why-global-temperatures-matter/.
2. Etan Siegal, "This Is Why Global Warming Is Responsible for Freezing Temperatures Across The US," Forbes, January 30, 2019. https://www.forbes.com/sites/startswithabang/2019/01/30/this-is-why-global-warming-is-responsible-for-freezing-temperatures-across-the-usa/#7d5a7a4cd8cf.
3. Sarah Gibbens, "Why cold weather doesn't mean climate change is fake," National Geographic," February 23, 2019. https://www.nationalgeographic.com/environment/2019/01/climate-change-colder-winters-global-warming-polar-vortex/.
4. Ellen Gray and Jessica Merzdorf, "Earth's Freshwater Future: Extremes of Flood and Drought" NASA, June 13, 2019. https://climate.nasa.gov/news/2881/earths-freshwater-future-extremes-of-flood-and-drought/.
5. Joseph Guzman, "Sea-level rise is accelerating along US coasts," The Hill, February 4, 2020. https://thehill.com/changing-america/sustainability/climate-change/481462-sea-level-rise-is-accelerating-along-us-coasts.
6. M Jackson, "Yes, glaciers are disappearing – but that's far from their only story," IDEAS.TED.COM, May 5, 2019.
https://ideas.ted.com/yes-glaciers-are-disappearing-but-thats-far-from-their-only-story/.
7. Peter Dockrill, *"Devastating Simulations Say Sea Ice Will Be Completely Gone in Arctic Summers by 2050,"* Science Alerts, April 23, 2020. https://www.sciencealert.com/arctic-sea-ice-could-vanish-in-the-summer-even-before-2050-new-simulations-predict.
8. Maddie Burakoff, "One Million Species at Risk of Extinction, Threatening Human Communities Around the World, UN Report Warns," Smithsonian Magazine, May 6, 2019.
9. Jeff Berardelli, "How climate change is making hurricanes more dangerous – Stronger wind speeds, more rain, and worsened storm surge add up to more potential destruction," Yale Climate Connections, July 8, 2019. https://www.yaleclimateconnections.org/2019/07/how-climate-change-is-making-hurricanes-more-dangerous/.
10. Emma Newburger, "Climate change is driving widespread forest death and creating shorter, younger trees," CNBC, Mat28, 2020. https://www.cnbc.com/2020/05/28/climate-change-is-driving-widespread-forest-death-creating-shorter-trees.html.
11. Jessie Yeung, "Climate change could kill all of Earth's coral reefs by 2100, scientists warn," CNN,

Fatal heat waves[12]
Skinnier pigs[13]
Fatter Horses[14]
Crop failure and food shortages[15]
Acidic oceans that will kill most marine life[16]
Billions of climate change refugees[17]
Increased cancer, cardiovascular disease, mental illness, and respiratory disease[18]
And, a devastating effect on the production of French wines[19]

The fact is there is no hard evidence that any of these things have been or will be triggered by human-caused emissions of CO_2. It is all conjecture based on the hypothesis that carbon dioxide controls temperature, which itself has never been determined as fact. More importantly most of these claims are predictions about things that haven't occurred to date and may never occur. In addition, many of these predictions are based on simulations, which are computer-generated models created by authors who decide what they want their model to predict and then build assumptions into the model that provide them with the results they are looking to achieve. It's all founded on a very self-fulfilling prophecy and has nothing to do with real science, which is about observing real situations in the real world, not inserting numbers designed to obey formulas in a computer. Public funding for computer-model predictions should be ended for both climate change and pandemic viruses

---

February 20, 2020. https://www.cnn.com/2020/02/20/world/coral-reefs-2100-intl-hnk-scli-scn/index.html.

12. Jacob Dubé, "As Canadian cities prepare for more deadly heat waves, limiting increase of climate change could save lives," *National Post*, June 13, 2019. https://nationalpost.com/news/world/as-canadian-cities-prepare-for-more-deadly-heat-waves-limiting-increase-of-climate-change-could-save-lives.

13. Amkur Paliwal, "A Warming Climate Could Make Pigs Produce Less Meat," *Scientific American*, September 24, 2018. https://www.scientificamerican.com/article/a-warming-climate-could-make-pigs-produce-less-meat/.

14. Brendon McFadden, "Climate change is making horses fat as it's causing an abundance of grass to grow, top vet warns," the *Telegraph*, November 29, 2019. https://www.telegraph.co.uk/news/2019/11/29/climate-change-making-horses-fat-causing-abundance-grass-grow/.

15. Christopher Flavelle, "Climate Change Threatens the World's Food Supply, United Nations Warns," the *New York Times*, August 8, 2019.

16. Alejandra Borunda, "Ocean acidification, explained," *National Geographic*, August 7, 2019. https://www.nationalgeographic.com/environment/oceans/critical-issues-ocean-acidification/.

17. Brian Palmer, "By 2070, More Than 3 Billion People May Live Outside the "Human Climate Niche," Natural Resources Defense Council, May 14, 2020. https://www.nrdc.org/stories/2070-more-3-billion-people-may-live-outside-human-climate-niche.

18. Isabella Annesi-Maesano, "The impacts of climate change on non-communicable diseases in the Mediterranean region," Sub-chapter 2.5.2., *The Meditteranean Region Under Climate Change*, OpenEdition Books, Marseille, 2016. https://books.openedition.org/irdeditions/23700?lang=en.

19. Rachel Tepper, "Climate Change May Drastically Reduce Traditional Wine-Producing Areas, Study Says" *Huffington Post*, December 6, 2017. https://www.huffingtonpost.ca/entry/climate-change-wine_n_3039673?ri18n=true.

alike. Billions of dollars are wasted on these programs. It's time everyone realized that computer models are not a crystal ball, which is after all a mythical object.

It is useful to know the difference between a skeptic and a heretic: a skeptic is someone who rejects the conclusions of a particular theory whereas a heretic rejects the underlying assumptions that are the basis for those conclusions. I'd much rather be labeled a climate skeptic or a climate heretic than a climate denier, which the not-so-nice people in the climate-catastrophe camp seem to prefer to use. The term "denier" is a silly moniker designed to culturally associate skeptical or heretical scientists with those who deny the reality of the Holocaust. But as silly and irrelevant as that association seems, that is what the purveyors of the fake, impending apocalypse have stooped to.

As for the world "getting too hot for life," the alarmists are neglecting to inform you of a well-known fact. When the Earth warms, as in the long "hot-house" eras of the past, it does not change much at all in the tropics. The warming occurs mainly towards the poles, thus raising the average global temperature while the tropics warm or cool very little, if at all. This results in a world where the difference in temperature between the tropics and the poles is far less than today during this present Pleistocene Ice Age.

Few people are familiar with the fact that humans are a tropical species as opposed to a sub-tropical, temperate, or polar species. Humans evolved in an equatorial environment where the temperature does not fluctuate much with the seasons and seldom goes below 20ºC (68ºF) at night or above 35ºC (95ºF) in the day. A few animal hides were all that was needed for cool nights. Even during the height of the glacial maximums, the temperature at the equator was very similar to today's while the temperate and polar regions were much colder. And even today, in this interglacial period, the temperate and polar regions of the planet are much colder than they were before the Earth cooled and settled into the Pleistocene Ice Age we are in now. As recently as three million years ago and before the onset of the Pleistocene Ice Age, giant camels roamed the Canadian high Arctic islands which were covered in forests (see Fig. 13).

The factors that made it possible for humans to migrate out of the tropics and eventually occupy most of the Earth's climates were fire, clothing, and shelter. Without these three essentials, humans could not even survive in the sub-tropics, or at high altitudes within the tropics. This is as true today as it was thousands of years ago.

Figure 13. Artist's depiction of giant camels that roamed the forests on Canada's Arctic islands such as Ellesmere Island above the Arctic Circle. Then the Pleistocene Ice Age settled in 2.6 million years ago and the camels and forests were gone.*

* Martha Harbison, "There Used to Be Freaking Camels In The Arctic," *Popular Science*, March 6, 2013. https://www.popsci.com/science/article/2013-03/there-used-to-be-freaking-camels-in-the-arctic/.

Consensus

One of the climate alarmists' more common strategies to avoid discussing specific points relating to the Earth's climate is to claim there is "an overwhelming consensus among scientists," and this is often placed at 97 percent. These alarmists claim there is a climate catastrophe approaching in the near future that will wipe out much of civilization and living creation if we continue to use fossil fuels at the rate we are presently using them. They ignore the fact that a consensus about something, by itself, gives us no clue whether the claim is correct or not. They also claim that "the science is settled," apparently in their favor. Other objectionable assertions they use are: "the science says (fill in the blanks)" and "it's simple physics." Science doesn't "say" things and physics isn't simple. Science is a process, not an oracle, and physics is highly complex and often, not even slightly intuitive.

It is a logical fallacy to use the assertion that most people agree with something as if that, in itself, proves something is true. "Consensus" is actually not a valid scientific term. It is a social and political term having to do with agreement on policies such as regulations, codes of conduct, procedures for making a decision, etc. Good policies are decisions, hopefully made democratically, that are based on proven facts.

Facts are determined by science. As the late Michael Crichton put it, "If it's consensus, it isn't science. If it's science, it isn't consensus."[20]

All through history brilliant scientists have been opposed by false consensuses. Galileo (astronomy), Mendel (genetics), Darwin (evolution), and Einstein (physics) each faced massive opposition to their discoveries. When Einstein, as a young patent clerk with no academic position published his *Theory of Relativity*, he was countered by a book titled *100 Authors Against Einstein*. Einstein's response to this, as he explained to a journalist, was: "If I were wrong, then one (author) would have been enough."[21]

The take home message here is that when someone begins a discussion of climate change with the claim that there is an overwhelming consensus, which itself is false,[22] listeners should probably change the subject. Mass delusion is also a consensus, as are cult movements and other such collective hysterias.

Carbon Dioxide

Those two words describe the most important molecule for the existence of life on Earth. That's because carbon dioxide is the source of the carbon for all carbon-based life, which of course represents all life on Earth. Add water, H_2O, to the equation and through photosynthesis green plants produce sugars such as glucose. These sugars provide the energy for the photosynthetic plants that produce them and for all the other species that cannot perform photosynthesis, including ourselves and all other animals. It really is that simple. But you can't observe this process because CO_2 is invisible, and the reactions involved in photosynthesis and energy-use are at the molecular level, inside the plants and animals, where observers cannot witness them directly.

I once laughed at people who said their houseplants grew better when they talked to them. How could anyone believe that? Plants don't have hearing. But then I learned that when we exhale there is 40,000 ppm (parts per million) of carbon dioxide in our breath. That is 100 times the amount of carbon dioxide there is in the air. When you

---

20. Michael Crichton, *State of Fear*, HarperCollins, 2004.
https://www.amazon.com/State-Fear-Michael-Crichton/dp/0066214130.
21. Stephen Hawking, *A Brief History of Time*, Bantam London, 1988.
https://www.amazon.com/Illustrated-Brief-History-Updated-Expanded/dp/055310374.
22. Anthony Watts, "Cooks '97 percent consensus' disproven by a new peer reviewed paper showing major math errors," September 3, 2013. https://wattsupwiththat.com/2013/09/03/cooks-97-consensus-disproven-by-a-new-paper-showing-major-math-errors/.

stand close and talk to your plants you are literally breathing concentrated fertilizer onto them.

Carbon dioxide originated from massive volcanic eruptions when the Earth was younger and much hotter. The heat that makes the Earth's core molten is mainly caused by radioactive decay of isotopes such as uranium, radon, thorium, and potassium. Some of these radioactive elements have decayed to much lower levels or even decayed altogether during the 4.6-billion-year history of the Earth. Therefore, the volume of carbon dioxide being produced by volcanic activity is now orders of magnitude lower than it was in Earth's earlier years.

The entire "climate change catastrophe" narrative is based on the claim that humans are emitting *too much* CO_2 into the atmosphere and that this will cause our planet to be *too hot* for life. Both of these claims are patently false.

The level of carbon dioxide in the atmosphere had been declining slowly and steadily for 150 million years, well before we began to use fossil fuels. During that period, it went down from about 2,000 to 2,500 parts per million (ppm) to 180 ppm during the peak of the most recent glaciation, 20,000 years ago. This was the lowest it has been in at least 500 million years, and probably since the Earth was formed 4.6 billion years ago. As the climate warmed out of the last glacial maximum during the 10,000 years it took to reach the warmer climate of the past 10,000 years, the warming oceans gave off CO_2 emissions and eventually brought atmospheric CO_2 up to 280 ppm. This is referred to as the "pre-industrial" level, around the year 1850, when carbon dioxide from burning fossil fuels for energy began to become largely responsible for increasing CO_2 to the present level of 415 ppm.

It is important to point out that even at 415 ppm, carbon dioxide is still much lower than it had been during the majority of the existence of modern life.[23] The chart below shows that CO_2 was at about 6,000 ppm at the time modern life emerged, 15 times higher than today (see Fig. 14). It then sank to about 500 ppm during the Carboniferous Period and eventually recovered to about 2,500 ppm during the Jurassic Period. It is not certain what factor(s) caused the decline of CO_2 in the Carboniferous Period. This period marked the advent of trees and forests which pulled a large amount of carbon dioxide from the atmosphere, but that in itself does not explain such a large decline in CO_2.

---

23. "Modern life" refers to the period named the Cambrian, when multi-celled life first evolved about 570 million years ago. Prior to that, for about three billion years, all life was unicellular, microscopic, and confined to the oceans and probably freshwater habitats as well. Life did not come on the land until about 440 million years ago.

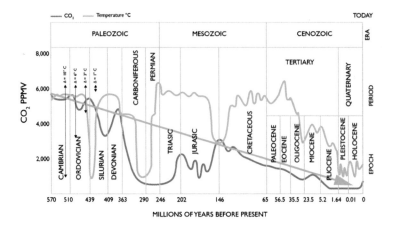

Figure 14. A 570-million-year record of carbon dioxide and global temperatures shows that while temperature rises and falls with no distinct trend, CO_2 has a net decline of more than 95 percent to the level of 280 ppm, before human emissions began to increase it. The little uptick at the far right indicates our recent contribution. Note there are periods of tens of millions of years when there is no correlation between carbon dioxide and temperature. This rules against a cause-effect relationship.*

* Nasif Nahle, "Cycles of Global Climate Change," *Biology Cabinet* journal Online, July 2009. http://www.biocab.org/Climate_Geologic_Timescale.html. Referencing C. R. Scotese, *Analysis of the Temperature Oscillations in Geological Eras*, 2002; W. F. Ruddiman, *Earth's Climate: Past and Future*, New York, NY: W.H. Freeman and Co., 2001; Mark Pagani, et al., "Marked Decline in Atmospheric Carbon Dioxide Concentrations during the Paleocene," *Science 309*, no. 5734 (2005): pp600-603.

Advocates of the climate-catastrophe narrative want us to ignore the record of carbon dioxide levels in the atmosphere before 1850 so they can make it look like 415 ppm is "high" compared to the pre-industrial level of 280 ppm. But the little uptick on the far right of the graph, indicating our contribution, shows that CO_2 levels are still at one of the lowest levels it has ever sunk to during the past 570 million years. The alarmists want the public to be conditioned into thinking that the past 170 years tells us everything we need to know about CO_2. As can be seen from the graphic above, there is much more to it than that.

The fundamental question is whether or not the claim that carbon dioxide is the "control knob" of global temperature is valid.[24] The chart above shows very clearly that CO_2 and temperature are out of sync more often than they are in sync. This does not support the claim that there

---

24. "How Carbon Dioxide Controls Earth's Temperature," NASA GISS, October 14, 2010. https://www.giss.nasa.gov/research/news/20101014/.

is a strong cause-effect relationship between CO_2 and temperature over the long-term history of the Earth; in fact, it rules against this conclusion. We are being told that the correlation between carbon dioxide and temperature both rising concurrently over that past 170 years, out of 570 million years of Earth's history, proves a cause-effect relationship. It does not, and the historical record indicates the opposite.

Figure 14 shows that before and after the 146-million-year mark, carbon dioxide and temperature were 100 percent out of sync with each other for about 30 million years before the mark, and for about 100 million years after. There are other long periods when they move in opposite directions, for instance during the Silurian Period and the Permian Period. The simultaneous rise of carbon dioxide and temperature over the last 170 years in no way supports a strong cause-effect relationship, in fact it is sufficient to reject that conclusion.

The relationship between "correlation" and "causation" is a subtle one, and the discernable difference between them is that correlation – two factors appearing to be strongly related – does not prove causation, whereas causation requires a strong correlation. For example, the correlation of two factors is often caused by a third factor. A very humorous example of this is the very high correlation between ice cream consumption and shark attacks. Obviously neither one of these factors cause the other. They are both caused by a third mutual factor, warm weather versus cold weather. In warm weather, people go to the beach, have an ice cream cone, and then go for a swim in the ocean and some get attacked by a shark. In the winter people are less likely to get an ice cream cone or go for a swim in the ocean, thus shark attacks are less frequent. Remember this example when you are told one thing causes another. There is an excellent six-minute TEDx Talk on the difference between correlation and causation.[25] Likewise, there is a hilarious website called Spurious Correlations that presents many strong correlations among two factors that are very obviously not in a cause-effect relationship.[26]

To conclude this section on the history of carbon dioxide, there is one question I have never heard an answer for. If CO_2 was 4,000 ppm at one time and 2,000 ppm at another time and still 1,000 ppm at another time and then during the Pleistocene rose and fell from about 190 ppm to 280 ppm numerous times, why is 280 ppm, the "pre-industrial" level considered some kind of benchmark level for

---

25. Ionica Smeets, "The danger of mixing up causality and correlation," November 5, 2012. https://www.youtube.com/watch?v=8B271L3NtAw.
26. "Spurious Correlations." https://tylervigen.com/old-version.html.

carbon dioxide in the atmosphere? Life on Earth worked just fine at all those other levels, which are clearly much higher. One climate activist group even named themselves 350.org to identify 350 ppm as the calibrated limit before bad things happen to the planet. Well, it's measuring at 415 ppm in the year 2020 and the main effect has been a big increase in the growth of plant life.

The Great Decline of CO_2 – Why did it Happen?

Looking back to Figure 14 above, it is clear during recent times, before we began emitting CO_2 from fossil fuel use, that carbon dioxide had actually declined to the lowest level it has ever been during the past half-billion years.

As previously mentioned, beginning about 150 million years ago, carbon dioxide has steadily declined to the lowest known level in the history of life on Earth. During the last glacial maximum, 20,000 years ago, CO_2 fell to about 180 ppm, only 30 ppm above the level where plants begin to die from CO_2 starvation. At lower than 150 ppm there simply isn't a high enough concentration of carbon dioxide for plants to survive. It's the same phenomena as animals suffocating from lack of oxygen. The graph below shows that if the decline in carbon dioxide continued along the same trajectory as it has been for all those millions of years, it would eventually sink to a level that would not support plant growth (see Fig. 15). This is not a very desirable outcome as it would threaten the survival of every living species on Earth. One might think this would have been noticed by those who claim there is "too much" CO_2.

The atmosphere contains 20.5 percent oxygen by volume, or 205,000 ppm, compared to carbon dioxide which is 0.0415 percent by volume, or 415 ppm. At six percent or 60,000 ppm oxygen, a human becomes unconscious and dies; whereas many plants can survive at 0.015 percent or 150 ppm carbon dioxide before they die from lack of it. It could be said that plants are (60,000/150 = 400) 400 times more efficient at obtaining their essential gas than humans. Amazingly some fish can survive at oxygen levels of two ppm in seawater, therefore, they are (60,000/2 = 30,000) 30,000 times more efficient at getting their essential dissolved gas than humans are at getting their essential atmospheric gas.

The cause of this eons-long decline in carbon dioxide has apparently not interested the proponents of climate disaster. This is presumably because they are trying to convince us that CO_2 is too high now and recognizing that it's at one of the lowest levels in Earth's history just doesn't support that narrative. But there is no question that this is

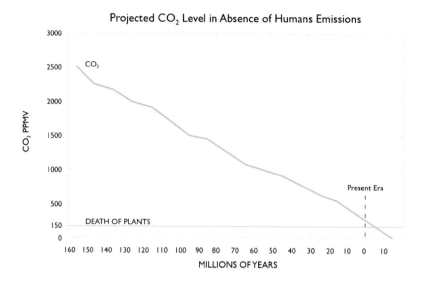

Figure 15. The steady decline in atmospheric CO_2 during the past 150 million years. During the last glacial maximum, 20,000 years ago, carbon dioxide sank to 180 ppm, only 30 ppm above the level where plants begin to die. Human-caused CO_2 emissions from fossil fuels and cement production have reversed that trend and restored a balance to the global carbon cycle. We are life's salvation, not its destroyers.

true. Marine sediment core samples contain this historical record and go back hundreds of millions of years and the sediment does not lie. Scientists who know their geological earth history agree with this data. The alarmists simply don't want to talk about this inconvenient truth.

One way the Earth's deep past is dismissed is with headlines proclaiming, "The Last Time CO_2 Was This High, Humans Didn't Exist."[27] This is true, but humans did not exist until about two million years ago. As carbon dioxide declined from 2,500 ppm during the past 150 million years, it was not until about five million years ago that it had declined to today's level of 415 ppm. It is rather preposterous to suggest that 415 ppm is somehow dangerous when all animal life, including our mammalian ancestors, lived through millions of years where carbon dioxide levels were at 2,000 ppm and higher.

The truth is, even at 415 ppm CO_2 most plants are limited in their growth rates and productivity because carbon dioxide is not at a sufficiently high level to achieve maximum growth. This is why virtually

---

27. Andrew Freedman, "The Last Time CO_2 Was This High, Humans Didn't Exist," Climate Central, May 3, 2013. https://www.climatecentral.org/news/the-last-time-co2-was-this-high-humans-didnt-exist-15938.

all commercial greenhouse growers pay for bottled CO_2 or purchase CO_2 generators that convert natural gas or propane into carbon dioxide. The optimum level of CO_2, taking into account both crop yield and optimum economics, is between 800 and 1,200 ppm for most greenhouse plants. In other words, plants thrive at two to three times the present level of carbon dioxide in the Earth's atmosphere. At 1,000 ppm photosynthesis is increased by 50 percent.[28] Higher levels of CO_2, up to 2,000 ppm, would promote even greater growth rates but there are diminishing returns above about 1,000 ppm so economics becomes the deciding factor due to the cost of carbon dioxide.

Another way of looking at this is that all farm crops grown in open fields, if they are receiving sufficient water and nutrients, are limited in their growth by a lack of carbon dioxide in the air. The same is true of all wild plants that have sufficient water, nutrients, and sunshine. They too are limited by today's historically low level of CO_2. Only if wild plants are growing in very arid regions, in soils lacking sufficient nutrients, and/or with insufficient sunshine does carbon dioxide not become the limiting factor for growth. Most of the world's plants, however, are starving for it.

So how did this situation come about where during the past 570 million years, carbon dioxide in the global atmosphere was reduced to less than four percent of its original level, from 6,000 ppm to under 200 ppm at its lowest during the past glaciation? The answer is quite surprising.

The fact that the level of carbon dioxide in the atmosphere has steadily declined during the past 150 million years clearly indicates there has continually been more CO_2 removed from the atmosphere than is being added to it during this very long time. Where did the missing CO_2 go, and more specifically where did the carbon go that was in that CO_2 in the atmosphere (see Fig. 16)?

The numbers are all expressed as petagrams (one petagram, also called a gigaton, equals one billion tons) of carbon. Some points of interest in the diagram are:

- Plants and soils together contain more than twice as much carbon as the entire global atmosphere. This was not the case when atmospheric CO_2 was five to ten times higher.

---

28. T. J. Blom, et al., "Carbon Dioxide in Greenhouses," Ontario Ministry of Agriculture and Rural Affairs, August 2009. http://www.omafra.gov.on.ca/english/crops/facts/00-077.htm

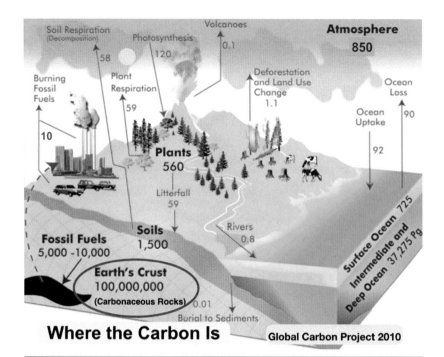

Figure 16. A graphic depiction of the flows (in red) and stocks (in blue) of carbon in the global carbon cycle as of 2020. Note that human emissions from fossil fuel use is estimated at 100 times the amount from volcanos, which were the original source of CO_2 in the atmosphere, much of which was absorbed by the oceans in earlier times. For the past 150 million years, volcanos have not emitted enough carbon dioxide to make up for the losses to sediments. Global Carbon Project, 2010* (updated to 2019 for fossil fuel emissions and atmospheric content).

* Jennifer LaPan, "Workshop Preps Educators to Train Next-Gen Carbon Researchers," NASA, May 12, 2013. https://www.nasa.gov/topics/nasalife/features/globe-workshop.html.

- The surface of the World Ocean, variously placed at 200 to 300 meters (656 – 984 feet) deep, contains nearly as much carbon dioxide as the global atmosphere, and the entire World Ocean contain 45 times as much CO_2 as the atmosphere. The vast majority of the carbon dioxide in the World Ocean was absorbed at the surface from the atmosphere, although a small amount was the result of underwater volcanic activity.
- The CO_2 absorbed by plants as they grow in the spring and summer is nearly balanced out on an annual basis, as plants emit carbon dioxide as they decay in fall and winter. The northern hemisphere dominates as there is much more land in the northern compared to the southern hemisphere.
- Since we began to emit CO_2 from fossil fuel use, especially since about 1950, plants have begun to grow faster and to

increase the global biomass each year. This means they are absorbing more carbon dioxide than they are emitting each year. This is known as "global greening" or the "CO_2 fertilization effect." This will be covered in another section of this chapter.

Fossil fuel reserves including coal, petroleum, and natural gas are estimated to contain between six and twelve times as much carbon as the atmosphere. The carbon in these hydrocarbons was removed from the atmosphere by plants on land and in the sea, and then got buried and was lost to the carbon cycle for millions of years. When we use fossil fuels, we are returning CO_2 to the atmosphere where it came from in the first place. Our emissions have brought an end to the 150 million-year decline of carbon dioxide and we are beginning to restore a balance to the global carbon cycle. There is simply no reason to believe CO_2 would not have continued to decline until there was too little for plants to survive. In this sense we are actually the saviors of life, not its destroyers as is claimed by the catastrophists.

But the truly astonishing number is that "carbonaceous rocks," such as limestone, marble, and chalk, contain 100 million billion tons of carbon. These minerals are composed of calcium carbonate, or $CaCO_3$, which is produced by marine calcifying species such as clams, corals, crabs, and microscopic plankton in order to produce armor in the form of shells to protect their soft bodies. $CaCO_3$ is produced by combining carbon dioxide and calcium dissolved in seawater, therefore removing carbon from the seawater.

The ability to control the crystallization of $CaCO_3$ is called "biomineralization" and it evolved in many diverse marine species at the beginning of the Cambrian Period 570 million years ago. These deposits contain 118,000 times as much carbon as there is in the Earth's atmosphere and 2,400 times as much carbon as in the atmosphere, oceans, plants, and soils combined. This latter carbon could be called the "free carbon" as it is actively moving through all these components on an annual or multi-annual cycle. The carbon in buried fossil fuels, and in particular in carbonaceous rocks, could be called "trapped carbon" as it is not available to living things, even though it once was millions of years ago, before it was turned into fossil fuels and rocks. Scientists refer to the trapped carbon as "sequestered carbon."

The diverse species that developed the ability to make calcareous shells for themselves include: the single-celled phytoplankton (coccolithophores), the single-celled zooplankton (foraminifera), the mollusks, corals, and all the crustaceans (crabs, shrimp, lobsters, and barnacles) (see Fig. 17).

These carbon-bearing rocks are not igneous rocks like granite and basalt that originated in the Earth's molten interior coming to the surface by volcanism. Most of these rocks were made by life itself, and they comprise a vast area of the surface of the Earth's crust. Anywhere there was once an ocean less than 1,880 meters (6,000 feet) deep during the past 500 million years – the Great Plains of North America for example – there are large limestone deposits. Some of the shale deposits that are fracked for oil and gas contain ten to sixty percent limestone, largely from the shells of calcifying marine species that sank to the ocean floor when they died. The oil and gas were transformed by heat deep in the sediments – from the soft parts of the living organisms that were inside the shells.

It is news to most people that we are at the tail-end of a 150-million-year decline of carbon dioxide in the atmosphere, caused primarily by multiple species of shelled-marine organisms consuming millions of billions of tons of CO_2 out of the ocean that resulted in the production of rocks. This is a testament to the sorry state of global education, the media, and many of our political and scientific institutions. Let's just say there is a lot of catching up to do in the "learning about life" department.

Figure 17. Here are some examples of the highly diverse marine calcifying species that have evolved the ability to produce protective armor for their soft bodies by combining carbon dioxide and calcium to make calcium carbonate. Clockwise from the left: the microscopic phytoplankton; coccolithophores (one of the most important primary producers in the oceans); mollusks such as clams, oysters, and snails; foraminifera that graze on blooms of coccolithophores and other phytoplankton; coral reefs which are responsible for about 50 percent of ocean calcification; and central in the figure, the crustaceans such as crabs, shrimp, and lobsters.

There are not many aspects of the CO_2 story that I am more certain of than this particular hypothesis regarding marine calcifying species; and the numbers speak for themselves. It is one "narrative" that is not based on computer models or the belief that everything that happened before 1850 is irrelevant. My first public presentation of this hypothesis was at the annual lecture for the Global Warming Policy Forum in London, England in October of 2016. I then published a peer-reviewed paper on the subject in March of 2017. Not one advocate for the "climate catastrophe" has attempted to rebut my thesis. It is simply ignored because they have no basis to dispute it.[29]

It is interesting to note that the decline in carbon dioxide was not done maliciously. It was a completely inadvertent result of many marine species evolving the ability to build armor plating for their soft bodies from the fusing of carbon dioxide and calcium. This evolution gave them a huge survival advantage. It is ironic though, that life itself developed a survival advantage that would eventually be a dire threat to life itself. In a similar manner, the human discovery of fossil fuels – fuels that became the source of 80 percent of our energy supply today – have inadvertently caused this decline in CO_2 to come to an end. This has the promise to bring CO_2 back to historic levels that are much more beneficial to nearly all plants and therefore, all life. But just because we didn't intend for this increase in carbon dioxide emissions doesn't mean we shouldn't recognize and celebrate one of the most positive developments in Earth's history, the replenishment of the most important substance for all life on Earth, after many millions of years of decline.

Burning fossil fuels for energy will stave off the chance of CO_2 decline for some centuries to come, but fossil fuels will eventually become scarce and civilization will likely turn to nuclear energy for the majority of its needs. It will not be necessary to emit as much carbon dioxide as we are today in order to retain a stable level in the atmosphere once it reaches a desirable level, of say 800 to 1,000 ppm. Nuclear energy does not produce CO_2 directly like fossil fuels, so it cannot replace fossil fuels as a direct source for carbon dioxide emissions for the sake of retaining an optimum level for life in the atmosphere.

This is where limestone and chalk come to the rescue. They will continue to be used for cement production and thus continue to produce a large volume of carbon dioxide. At the present, cement production accounts for five percent of human-created CO_2 emissions. It

---

29. Patrick Moore, "The Positive Impact of CO_2 Emissions on the Survival of Life on Earth," Frontier Centre for Public Policy, March 2017. https://www.dropbox.com/s/uhq557vrnww0ala/PositiveImpactOfCO2ForLife.pdf?dl=0.

is possible to use both nuclear energy, for which there is a resource of fuel for many thousands of years, and direct solar energy to convert limestone into lime and CO_2.[30] By returning some of the carbon dioxide that plants and marine species removed from the atmosphere, and will continue to remove into the foreseeable future, humans can be the keepers of life for millennia to come. Perhaps one day in the distant future every country will join a treaty that sets a quota for their annual requirement of CO_2 emissions. We should celebrate carbon dioxide.

The Greenhouse Effect

The greenhouse effect is another invisible factor, and like CO_2, it is definitely a very important one. It is also a good reminder that physics is anything but basic or simple. While carbon dioxide is necessary for all life on Earth, the greenhouse effect, caused by greenhouse gases in the atmosphere, makes the Earth warm enough for life in the first place. This is because the nature of the greenhouse effect is to impede the rate that radiation (heat), coming in from the Sun, escapes from the Earth and returns back into space.

The primary greenhouse gases are water vapor, carbon dioxide, and methane. Water vapor is estimated to be between 65 to 90 percent of the greenhouse effect, depending on the relative quantities of these gases at a particular location. Some climate websites simply fail to mention that water vapor (humidity) is the main greenhouse gas in the atmosphere.[31] Water vapor is also invisible.

Some people are critical of using the metaphor "greenhouse effect" for this phenomenon, as they point out there is no glass ceiling on the atmosphere like there is in a greenhouse. The warming in a greenhouse is mostly due to the glass ceiling preventing convection from taking the heat that came in from the Sun upwards and out of the greenhouse. The greenhouse effect caused by H_2O vapor, CO_2, and CH_4 is due to the re-radiation of heat back down to the ground by these atmospheric gases, thus slowing the cooling of the Earth's surface. Therefore "greenhouse effect" is not actually a very good metaphor for the effect of these gases, but that is the operative phrase used in the literature on climate change. These gases do result in reduced cooling, but by an entirely different mechanism than a greenhouse does.

---

30. Rajaram Swaminathan, "Design of Solar Lime Kiln," Innovative Energy & Research, September 25, 2017. https://www.omicsonline.org/open-access/design-of-solar-lime-kiln.php?aid=95624.
31. Land Trust Alliance, "Carbon Dioxide, Methane, Nitrous Oxide, and the Greenhouse Effect," Source: EPA, Undated. https://climatechange.lta.org/get-started/learn/co2-methane-greenhouse-effect/.

However, there really is a kind of ceiling in the atmosphere. It is at the top of the troposphere, the bottom layer of the atmosphere where life exists. The boundary between the troposphere and the next layer, the stratosphere, is named the tropopause (see Fig. 18). The tropopause is a strong barrier to convection currents that carry heat upwards from the surface; where convection is essentially blocked by an analogous, but different mechanism than the glass that blocks convection in a greenhouse. The tropopause is the strongest inversion in the Earth's atmosphere, and like inversions in the lower atmosphere, it prevents upward flow of air and clouds, and the heat that is held in them. This phenomenon can be observed with the naked eye as thunderclouds form and carry heat skyward. When the clouds meet the tropopause, they flatten out at the top forming "anvil clouds" (cumulonimbus) (see Fig. 18).

The intersection between the troposphere and stratosphere is called the tropopause. In the stratosphere, the temperature increases with height. This is due to the production of heat in the process of the formation of ozone. The positioning of warmer air over colder air here is called a temperature inversion. Its effects are most obvious by the formation of anvil tops in cumulonimbus clouds, as the inversion prevents convection at the top of the troposphere (see Fig. 19).[32]

There are three primary mechanisms whereby heat moves from one place to another in the atmosphere: convection, conduction, and radiation. Convection effectively ends at the tropopause. Conduction is not a major factor in gases and effectively ends as the atmosphere thins with altitude. Therefore, the only mechanism whereby heat eventually escapes back to space, where it originally came from, is by radiation. Interestingly, whereas CO_2 is a comparatively minor greenhouse gas compared to H_2O near the Earth's surface, it is a primary molecule responsible for radiating heat back out of the atmosphere. There is very little H_2O above the tropopause, so CO_2 plays the largest role in cooling the Earth to balance the heat coming in from the Sun. Most of what we hear from activists, mainstream media, many politicians, and "climate scientists"[33] is that carbon dioxide causes dangerous warming, which is

---

32. University of Maryland, Department of Geology, Geology 100 Curriculum, April 2010. https://stratusdeck.co.uk/vertical-temperature-structure.
33. The term "climate scientist" is a bit of a trick. The study of climate involves a large number of disciplines, including atmospheric physics, oceanography, geology, paleontology, evolutionary biology, astrophysics, meteorology, and all aspects of biology (life science). The alarmist camp ascribes the term "climate scientist" to itself and those who agree with them, almost regardless of their core education. So, it turns out the opposite of a climate scientist is a "climate denier" and not worthy of interest. This is part of why they say 97 percent of "climate scientists" agree with them. It's because they think they can decide who is a climate scientist and who is not. This is not how science is supposed to work.

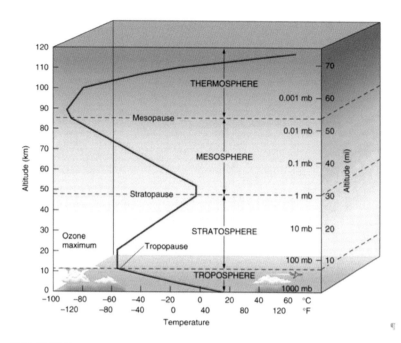

Figure 18. A graphic showing the layers in the atmosphere. Virtually all the weather is confined to the troposphere which has an average altitude of 11 kilometers (6.8 miles). The tropopause is a global inversion that stops convection. Above the tropopause the only way heat gets back to space is by radiation, mainly through carbon dioxide.

Figure 19. A classic example of convection ending at the tropopause due to the temperature inversion that blocks air from rising any further, thus not allowing the heat contained in the air to rise any further either.

not actually proven. While it is certain that carbon dioxide is a major player in returning the Sun's heat to space and balancing the temperature, and this is an entirely positive function.

In order for the Earth to remain at a stable temperature, the same amount of heat must be expelled from the Earth to space as the amount that radiates in from the Sun. When the Earth's climate is cooling, it is because more heat is escaping back into space than is coming in from the Sun, and conversely, when the Earth is warming that means more heat is coming into the Earth than is escaping back to space. This does not occur instantaneously as the oceans heat and cool much more slowly than the land and/or the atmosphere. This creates variable lag times in the Earth's response to incoming solar radiation and makes it very difficult, likely impossible, to predict or model future trends accurately.

Most of the Sun's heat is short-wave radiation that easily penetrates the atmosphere where there are no clouds, and even when there are clouds some heat still gets through to the Earth's surface. But because the Earth is so much cooler than the Sun, the radiation of heat going back to space is long-wave radiation (infrared or IR) which is invisible, much of which is intercepted by the greenhouse gases and re-radiated in all directions, including downwards towards the ground or water body where it came from. This is repeated many times as the IR energy makes its way back to space, continually being absorbed and re-emitted by the greenhouse gases. This slows the release of the infrared energy to space, making the atmosphere warmer than it would be in the absence of the greenhouse gases.

The greenhouse effect slows the cooling of the Earth, thus raising the temperature about 33ºC (59.4ºF) above what it would be if there were no greenhouse gases.[34] At the present time the average temperature of the Earth is about 14.9ºC (58.8ºF), which is not really that warm by any human standard. Imagine if there were no greenhouse effect and the average temperature was -18ºC (-0.5ºF). Life would likely never have existed without the greenhouse effect. And yet the climate alarmists refer to greenhouse gases such as CO_2 and CH_4 as "pollution." The greenhouse gases are, in fact, one of the main reasons for the existence of life on our planet.

One of the reasons it is possible for the alarmists to exaggerate the greenhouse effects of CO_2 and CH_4 as opposed to H_2O vapor, is the fact that it is actually impossible to directly measure the precise amount of

---

34. 33°C seems like a very large temperature increase, but when expressed on the Kelvin Scale, where absolute zero K is -273°C, it is actually a rise from 255°K to 288°K, or an increase of about 13 percent.

temperature increase that is caused by each of these individual greenhouse gases. This is partly because the wavelengths they re-radiate overlap in some cases and because the concentration of water vapor is constantly changing with both location and time.[35] Regardless of this limitation it is very clear that H_2O is by far the dominant greenhouse gas.

It is difficult to get an unbiased view of the relative importance of these three greenhouse gases. Some actually say H_2O vapor by itself has no effect on temperature even though it is much more abundant in the atmosphere than the other greenhouse gases. It is even claimed, with confidence, that water vapor will have a "positive feedback effect" on temperature, as carbon dioxide and methane climb higher leading to higher temperatures, which in turn will result in more water vapor (humidity) in the atmosphere. However, it is equally possible that the effect of higher temperatures will cause more water in the form of clouds, thus providing a "negative feedback effect" and reversing some or all of the warmth caused by CO_2 and CH_4. Today, clouds reflect nearly 25 percent of all sunlight back to space, thus making the Earth cooler than it would be with no clouds. But if the percentage of clouds increased due to warming, this could cancel out some or all of the effect of additional carbon dioxide. Clouds can be best described as "wild cards" when attempting to predict future effects of more CO_2 in the atmosphere. Imagine trying to construct a computer model that could predict the patterns and extent of global cloudiness on any given day 20 years from now (see Fig. 20, page 52). Joni Mitchell, a fellow Canadian, wrote these wise words for her song "Both Sides Now":

> *I've looked at clouds from both sides now*
> *From up and down and still somehow*
> *It's cloud's illusions I recall*
> *I really don't know clouds at all.*[36]

> *The climate system is a coupled non-linear chaotic system, and therefore the long-term prediction of future climate states is not possible.*
> Intergovernmental Panel on Climate Change – May 2018[37]

---

35. Thayer Watkins, "Saturation, Nonlinearity and Overlap in the Radiative Efficiencies of Greenhouse Gases," San Jose University, not dated. This is a good explanation of H_2O and CO_2 as greenhouse gases. https://www.sjsu.edu/faculty/watkins/radiativeff.htm.
36. Joni Mitchell, "Both Sides Now," Gandalf Publishing Co., 1967. https://jonimitchell.com/music/song.cfm?id=83
37. B. Moore, et al., "Advancing our Understanding" IPCC TAR-4, p774, March 2018. https://www.ipcc.ch/site/assets/uploads/2018/03/TAR-14.pdf.

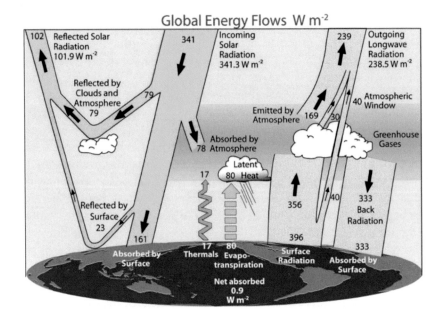

Figure 20. The various flows of incoming solar radiation and outgoing longwave (IR) radiation are known, however, accurately quantifying them all is not a simple matter. Most of them are invisible, and instruments are required to measure them. Note that back radiation from greenhouses gases in the atmosphere is nearly as large as the total amount of incoming solar radiation. Back radiation, here, is effectively magnifying the impact of solar radiation and raising Earth's temperature by 33°C. Also note the sum of incoming radiation absorbed by the atmosphere and adsorbed by the surface is equal to the amount of outgoing longwave radiation, 235 W/m2. This would be a state where no net warming or cooling are taking place, which would occur every time warming switches to cooling or vice versa.*
Alarmists would have you believe that the Earth will never cool again, when it is already cooler than it has been during most of life's existence. It is simple to see from this diagram that predicting future climates is not only daunting, but very likely impossible (After Trenberth, et al.**).

* J. T. Kiehl and K. E. Trenberth, (1997), "Earth's Annual Global Mean Energy Budget" Bulletin of the American Meteorological Association, 78: pp197-208. https://journals.ametsoc.org/bams/article/78/2/197/55482/Earth-s-Annual-Global-Mean-Energy-Budget.

** K. E. Trenberth, et al., "Estimates of the global water budget and its annual cycle using observational and model data," *Journal of Hydrometeorology*, 8:758–769, 2009. https://journals.ametsoc.org/bams/article/90/3/311/59479/Earth-s-Global-Energy-Budget.

The above quote from the IPCC is very clear, yet billions of dollars have been spent on climate models, nearly all of which have proven to greatly exaggerate what has actually occured. These funds would be more productively spent on climate change itself rather than on computer models that simply cannot predict the future global climate because of the factors listed in the IPCC quote above.

Prediction is very difficult, especially about the future.
—Niels Bohr – 1922 Nobel Prize for Physics

Carbon Dioxide and the Greening of the Earth

On one hand, carbon dioxide has been branded a "pollutant" by alarmists and even legally designated as such by the US Environmental Protection Agency under President Obama. Many other countries are treating it as if it is a large negative factor for climate and civilization. On the other hand, CO_2 is the staff of life, or the stuff of life, and it is proving to be an extremely positive factor for plant life as we increase its concentration in the global atmosphere. This is the primary demarcation in the debate about increasing carbon dioxide. Is it entirely negative? Is it partly negative and partly positive? Or is it entirely positive. I fall into the latter camp of thinking and I believe I can defend that position with confidence.

The "greening of the Earth" due to our CO_2 emissions, caused by an increase in the atmospheric level of carbon dioxide, has been known for some time. There is the unproven hypothesis, based on only 170 years of climate change, that claims CO_2 is a "control knob" and one that is causing global warming, and is entirely negative. To the contrary, the "global greening" effect of carbon dioxide is, without a doubt, proven to be true and is based on real measurements rather than computer models, false consensuses, and contrived "narratives."

One of the best experimental demonstrations of the increase in plant growth caused by additional CO_2 was conducted by Sherwood Idso, the founder of the website CO2science.org (see Fig. 21, page 54).

Figure 21. All four trees were grown under the same conditions except for the concentration of CO_2 in their plastic enclosures. This is why the Earth is greening as we elevate carbon dioxide in the atmosphere by nearly 50 percent, from a starvation-level of 280 ppm to 415 ppm. It can be seen from this experiment that there is room for much more growth in trees, food crops, and other plants as CO_2 continues to rise to more optimum levels. The level of 835 ppm of carbon dioxide allows trees to grow more than double the rate of 385 ppm. This photo was taken in 2009 when atmospheric CO_2 was about 385 ppm.

The first science agency to announce that the Earth was "greening" (increasing in plant biomass) was the Commonwealth Science and Industry Research Organization (CSIRO) in Australia in 2013 (see Fig. 22).[38] This is also confirmed by NASA (see Fig. 23).[39] A recent collaboration has determined that the additional, human-induced CO_2 emissions into the atmosphere has caused a global increase in biomass of 31±4 percent, nearly twice the previous estimates.[40]

Even if one accepts that carbon dioxide is responsible for most of the 1.0°C rise in the average global temperature during the past 170 years – which is a good thing no matter what caused it – the effect of increased CO_2 on the growth of food crops, trees, and many wild environments is by far the more beneficial impact. Yet alarmists ignore, downplay, and even condemn the greening as in this ridiculous offering from *New Scientist*:

---

38. Andrew Wright, "Deserts 'greening' from rising CO_2," CSIRO, July 3, 2013. https://csiropedia.csiro.au/deserts-greening-from-rising-co2/.
39. Sampson Reiny, "Carbon Dioxide Fertilization Greening Earth, Study Finds," NASA, March 27, 2019. https://www.nasa.gov/feature/goddard/2016/carbon-dioxide-fertilization-greening-earth.
40. Vanessa Haverd, et al., "Higher than expected CO_2 fertilization inferred from leaf to global observations," *Global Change Biology*, February 24, 2020. https://onlinelibrary.wiley.com/doi/full/10.1111/gcb.14950.

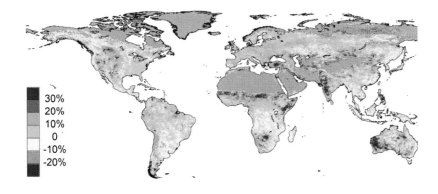

Figure 22. Depiction of the degree of greening that has occurred across the Earth between 1982 and 2010, from the Commonwealth Science and Industry Research Organization in Australia. As CO_2 continues to rise in the atmosphere, we can expect a continued increase in the global biomass of plants.

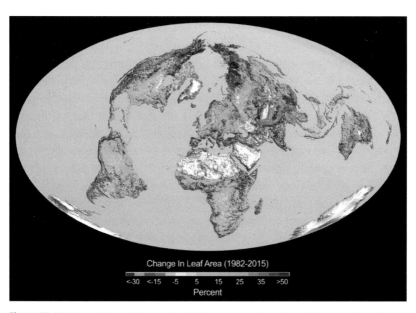

Figure 23. NASA's depiction of the greening Earth showing vast regions with increased growth due mainly to the higher atmospheric level of carbon dioxide caused by human emissions. A lusher world is a world richer in life.

Not so fast. According to Princeton biologist Irakli Loladze, we may have overlooked a potentially devastating consequence of rising CO_2 levels. It might be a boon to plants, but higher levels of the gas could trigger a pandemic of human malnutrition.

> At first, this sounds nonsensical. Surely faster-growing plants can only make food more plentiful? Indeed, it will, but quantity isn't the issue here. Loladze reckons we should be deeply worried about the quality of food from these plants. According to his analysis, crops that grow in high CO_2 are nutritionally barren, denuded of vital micronutrients such as iron, zinc, selenium, and chromium. If he's right, we're heading for a world where there's food, food everywhere, yet not a thing to eat.[41]

This is unadulterated fake news from a "science" magazine that long ago went tabloid, not unlike what happened with *National Geographic* more than 20 years ago. Any plant that was "denuded of vital micronutrients" would die quickly due to a lack of vital nutrients. As with humans, all plants and animals have a need for essential nutrients. It is true that when additional carbon dioxide causes plants to grow faster they need more of the other nutrients. So, give them more nutrients, like any good farmer would do. That is what's in "fertilizer." If you deprive plants of a particular essential nutrient, nitrogen for example, then that nutrient becomes the limiting factor for growth. The concept of "limiting factors" is one of the most important principles in biology. Essential minerals, like iron, and essential nutrients, like beta-carotene, must be provided to the plant or animal, otherwise that plant or animal will not grow. If you provide every essential nutrient in excess except for one, that lacking nutrient becomes the limiting factor for growth.

But there is an easier way to explain why this claim is fake. If it were true, then all the food produced in commercial greenhouses would be "nutritionally barren." Earlier I pointed out that it is a standard operating procedure for greenhouse growers to increase the level of CO_2 to 800 – 1,200 ppm to get an optimum yield. That is double or triple the present level of atmospheric carbon dioxide. In this case, there is no loss of nutritional value in greenhouse-grown crops because the grower increases the level of nutrients provided so the plants can grow faster and yield more food. Here is what a professional nutritionist has to say about greenhouse crops grown in hydroponics, without soil.

> It's also impossible to generalize about whether hydroponically grown veggies are more or less nutritious than conventionally grown. Part of this, again, will depend on the varieties being grown. And part of it depends on the growing medium. By increasing the concentration of nutrients in the hydroponic medium,

---

41. Graham Lawton, "Plague of Plenty," *New Scientist*, November 30, 2002. https://www.newscientist.com/article/mg17623715-200-plague-of-plenty/#ixzz6NaEEtfPy.

> *you can actually increase the nutrient content of the vegetable (my emphasis).*
>
> Whether conventional or hydroponic, many of those nutrients will begin to fade once the produce has been harvested. I've said this before about organic produce and it is equally applicable to hydroponics: How fresh the produce is may have more of an impact on its nutrient quality than how it was grown. Lettuce or tomatoes from a local hydroponic grower may be more nutritious than conventional or organic produce that's spent a week in transit, simply because less time has elapsed.[42]

So, the truth is that the nutritional value of foods grown in greenhouses, even without soil, can actually contain enhanced nutritional value if given sufficient nutrients. Today, farmers who grow their crops in open air are getting the benefit of a nearly 50 percent increase in CO_2 from the past 150 years which translates into about a 30 percent increase in yield from their crops. This is hugely positive for farmers and consumers.

Faster growth is not the only benefit of more carbon dioxide in the air. Higher levels of CO_2 also make plants more efficient in their use of water. When carbon dioxide was at about 280 ppm in 1850, plants had to work hard to get enough CO_2 for growth. Such low levels cause plants to produce more pores, called stomata, on their leaves where they take in carbon dioxide from the air. This is also where water is lost from the plant, through a process known as transpiration. When CO_2 is higher, plants produce fewer stomata and close these pores more often. This results in less water loss and, accordingly, plants can live where it was previously too dry for them when the carbon dioxide level was lower. In many parts of the world today, where it was too dry for trees before, they are now marching out into the grasslands and changing it into open forests.[43] Of course, this is also seen as a negative development by alarmists.[44] You get the impression they are simply against change, period; well, except that they want us to give up 80 percent of

---

42. Monica Reinagel, "Are Hydroponic Vegetables Less Nutritious?" Quick and Dirty Tricks, April 26, 2016. https://www.quickanddirtytips.com/health-fitness/healthy-eating/ask-the-diva/are-hydroponic-vegetables-less-nutritious.
43. Zender Venter, "Woody plants on the march: trees and shrubs are encroaching across Africa," *The Conversation*, August 13, 2018. https://theconversation.com/woody-plants-on-the-march-trees-and-shrubs-are-encroaching-across-africa-101135.
44. Arizona State University, "Trees, shrubs invading critical grasslands, diminish cattle production," *Science Daily*, August 18, 2014. https://www.sciencedaily.com/releases/2014/08/140818161355.htm. https://www.sciencedaily.com/releases/2014/08/140818161355.htm.

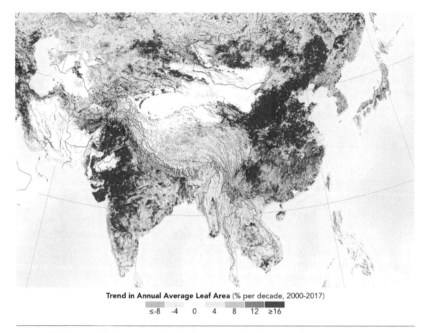

Figure 24. China and India account for the largest contributions to global greening. This is due to a combination of tree planting, intensive agriculture, and the CO_2-fertilization effect. These are positive trends, and the rise in global temperature by about 1.0°C during the past 300 years is also a positive trend.

our total energy supply – currently provided by reliable, cost-effective fossil fuels – and try to replace it with unreliable expensive forms of energy like wind and solar. Now that is a proposed change worthy of opposition.

One of the more interesting aspects of the greening trend is that the world's two most populous countries, China and India, are making the greatest contribution to the increase in global biomass, mainly trees and food crops.[45] In China this is due to a massive tree planting program on lands that were previously deforested; and in both countries it is partly due to intensive agricultural production to feed their large populations. Of course, the rising level of CO_2 will magnify the effect of all this human activity by making most plants and trees grow faster than before (see Fig. 24).

---

45. Abby Tabor, "Human Activity in China and India Dominates the Greening of Earth, NASA Study Shows," NASA, February 11, 2019. https://www.nasa.gov/feature/ames/human-activity-in-china-and-india-dominates-the-greening-of-earth-nasa-study-shows.

The Pleistocene Ice Age – The Great Cooling of the Earth

It is very important to get a perspective on how long life has existed and where we are now in the climate history of the Earth. Anyone who has studied this seriously knows that we are living in a very cold period when compared to the long-term history of the world's climate. We are currently in the Pleistocene Ice Age, but most climate alarmists want you to believe the Pleistocene Ice Age has ended and the Earth will now heat up catastrophically.[46] There is no justification for this fiction.

The Pleistocene Ice Age officially began 2.6 million years ago after a 50-million-year cooling period from the Eocene Thermal Maximum. Notably, 2.6 million years is a relatively short period in geological time, whereas tens of millions of years are considered serious geological time. There have been 460 tens of millions of years since the Earth was formed 4.6 billion years ago, and at least 350 tens of millions of years since life began. Contrary to the climate crisis narrative, we are still in the Pleistocene Ice Age today and there is absolutely no telling when it will end. If you need proof of this take note that both the Arctic and Antarctic are covered in massive glaciers and sheets of frozen ocean. True, we are in one of many relatively short interglacials of the Pleistocene, but even so, we are still living during record-low temperatures compared to most of Earth's history.

Thirty years ago, we were told by many pundits that the polar ice caps would be melted away long before now. It hasn't happened, and it is most certainly not going to happen anytime soon. Later in this chapter we will go back even further in time, but for now it is important to realize that during the past 50 million years the Earth has been cooling quite steadily. This is the total opposite of what you are being told by the alarmists who are declaring this a "climate emergency" or "climate crisis," and predicting an end to life as we know it if we continue along the present path of slight warming. The irony of what the alarmists are saying concerning the temperature of the Earth and that it is too hot, is that it is actually colder than it has been during most of life's existence, and that life, historically, has better flourished during the warmer periods than the comparatively colder periods, like we are in today.

The graph below was prepared in 2008 by James Hansen who is one of the most notorious climate alarmists.[47] I may never understand how

---

46. Ron Brackett, "World's Annual Temperature Could Hit 2.7-Degree-Rise Threshold Within Next Five Years, WMO Says," The Weather Channel, July 9, 2020. https://weather.com/science/environment/news/2020-07-09-annual-global-temperature-dangerous-rise-wmo.
47. James Hansen, "Storms of my Grandchildren, The Truth About the Coming Climate Catastrophe and

he could be aware of this temperature record and still be preaching that the Earth is heading for a global-warming catastrophe. But the graph itself is very understandable (see Fig. 25). Life on Earth is at the tail end of a 50-million-year period of global cooling, from a time when the Earth was ice-free for millions of years to today where high mountain glaciers are common and both poles are covered in ice year-round.

Antarctic glaciation began about 33 million years ago, as the southern hemisphere cooled millions of years earlier than the northern hemisphere. This is largely due to the fact that there is far less land and far more ocean in the southern hemisphere than in the northern hemisphere. Then there were eight million years of slight warming followed by a leveling off for another 10 million years, and finally the 12-million-year plunge into the Pleistocene Ice Age.

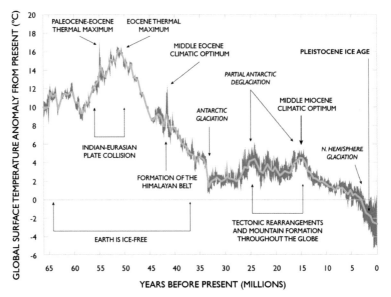

Dynamics of global surface temperature during the Cenozoic Era reconstructed from ^{18}O proxies in marine sediments (Hansen et al. 2008)

Figure 25. Until Antarctic glaciation began 33 million years ago, the Earth had been completely ice-free for more than 200 million years, since the end of the Karoo Ice Age. The present Pleistocene Ice Age is considered to have begun when the Arctic began to freeze about 2.6 million years ago. the very slight warming of 1.2°C since the year 1850 to 1.2°C (2.2°F), is inconsequential compared to the long history of the planet (After Hansen, et al., 2008).*

* Hansen, et al., "Target Atmospheric CO_2: Where Should Humanity Aim?, "The Open Atmosphere Science Journal," 2008, 2, pp217-231. https://pubs.giss.nasa.gov/docs/2008/2008_Hansen_ha00410c.pdf.

Our Last Chance to Save Humanity," *Bloomsbury USA*, December 21, 2010. https://www.indiebound.org/book/9781608195022.

Consider that Antarctica, which is considerably colder than the Arctic, began to freeze over when the average global temperature was about 6°C (11°F) warmer than it is today. There is no way a 2°C (3.5°F) rise in temperature will melt Antarctica. It is surrounded by a circumpolar cold current that keeps the warm water from the north away from the continent. It is possible there will be additional melting in the Arctic, but it is virtually impossible that it will become ice-free in the coming centuries.

The last time the Earth was as cold as it has been during this Pleistocene Ice Age was near the end of the Karoo Ice Age about 260 million years ago. During the intervening period, with the exception of a minor cooling 145 million years ago, the Earth has been in the Hothouse Ages, where there was no ice at the poles and the land at both poles was forested and warm. The Karoo Ice Age lasted for about 100 million years, from 360 to 260 million years ago. There is no guarantee that the Pleistocene Ice Age will not last that long too. We simply cannot predict the future climate when it comes to Ice Ages. They do not have a regular historical pattern. It is thought by some that the movement of the tectonic plates in the Earth's crust could change ocean currents – thus changing patterns of heat distribution – could be the cause, but this is not at all certain.

The most recent glacial maximum in the Pleistocene stretched across the entire North American continent to south of the present-day border between the United States and Canada (see Fig. 26, page 62). This ice sheet was about the same size as the Antarctic ice sheet today. During this glaciation, which peaked 20,000 years ago, present day Boston was under 1.25 kilometers (4,100 feet) of ice and Montreal was under 3.3 kilometers (10,560 feet) of ice (see Fig. 27, page 62). The change in global climate from then to the beginning of this interglacial, called the Holocene, makes our present situation a total non-event. The sea level rose 120 meters (395 feet) until all the low-elevation, mid-latitude glaciers were melted about 7,000 years ago. Since then the sea level has oscillated a few feet up and down following the relatively minor temperature fluctuations during the past 7,000 years of the Holocene Interglacial (see Fig. 29, page 63). The change from the glacial maximum to this interglacial was some real climate change. What has occurred with the 1.2°C (2.2° F) rise in temperature since 1850 is simply part of the minor, and quite normal, ups and downs during this interglacial.

Figure 26. Depiction of the ice sheets that covered nearly all of Canada and parts of all the northern-tier US states at the height of the last glaciation 20,000 years ago. The sea was 120 meters (395 feet) lower at that time, exposing the Bering Land Bridge that allowed humans to occupy the New World for the first time (see Fig. 28).

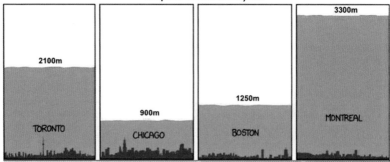

Figure 27. Depiction of the thickness of the ice sheets over the present locations of four North American cities 20,000 years ago. The skylines of the cities today give a good reminder of what real climate change looks like.

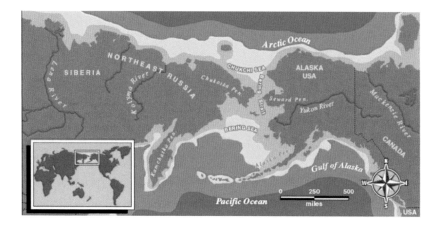

Figure 28. Map of the Bering Land Bridge during the last glacial maximum 20,000 years ago. Not only humans but reindeer (caribou), timber wolves, moose, and brown bears (grizzly bears) migrated to the New World from the Old World at that time. Horses and camels, which were New World species went the opposite way.

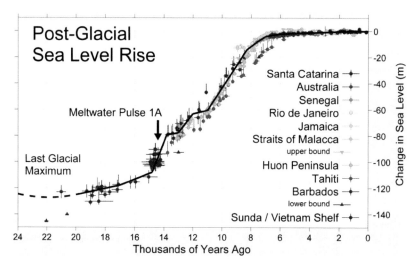

Figure 29. The 120-meter (395-foot) rise in global sea level coming out of the most recent glacial maximum. During the past 7,000 years the sea level has fluctuated in concert with the many minor ups and downs in global temperature.

Many people believe that the Pleistocene Ice Age is over, now that the ice that once covered more than half of North America 20,000 years ago has melted and we have been in a warmer interglacial period climate for the past 10,000 years. This belief is caused by confusing what is properly referred to as the most recent major glacial advance with the Pleistocene Ice Age itself. The most recent glacial advance was just the most recent of as many as 45 such advances during the Pleistocene Ice Age, which the Earth has been in for 2.6 million years. The Pleistocene Ice Age is, in fact, alive and well. So please take note that "last glacial advance" does not mean the "final glacial advance," it just means the "most recent glacial advance."

It doesn't help much that many so-called authorities on climate insist that the Pleistocene Ice Age is over. They purposely confuse "ice age" with "glacial period" as in this quote from the University of California Museum of Paleontology.[48]

> *The Holocene is the name given to the last 11,700 years of the Earth's history – the time since the end of the last major glacial epoch, or "ice age." Since then, there have been small-scale climate shifts, but in general, the Holocene has been a relatively warm period in between ice ages.*

First, they refer to the "last major glacial epoch, or 'ice age.'" It is impossible to know whether they are referring to the entire ice age, or only to the most recent major glaciation within the Pleistocene Ice Age. Then they state that "The Holocene has been a relatively warm period between ice ages," clearly inferring that another "ice age" is expected after this interglacial we are in today. This makes no sense as this Holocene Interglacial is no different in any fundamental way from the previous, approximately, 44 interglacials. Like the Holocene, the past few interglacials have been named. The authors also use the word "epoch" incorrectly when referring to the "glacial epoch." The Pleistocene Ice Age is an epoch, but the glacial advances during it are not. They are cycles within the Pleistocene Ice Age.

Climate alarmists seldom, if ever, mention the brilliant discovery made by Milutin Milankovitch, a Serbian astronomer, in the 1920s.[49,50] He determined that the gravitational effect of the position of the large

---

48. The Holocene Epoch, University of California Museum of Paleontology, June 10, 2011. https://ucmp.berkeley.edu/quaternary/holocene.php.
49. Milankovitch Cycles, https://en.wikipedia.org/wiki/Milankovitch_cycles.
50. https://climate.nasa.gov/news/2948/milankovitch-orbital-cycles-and-their-role-in-earths-climate/.

gas planets, Jupiter and Saturn in particular, cause regular cycles in three key factors of the Earth's pattern of movements.

- The 100,000-year cycle of the variation in Earth's orbit around the Sun. The orbit changes from more elliptical to less elliptical, thus resulting in varying distances from the Sun.
- The tilt, or "obliquity" of the Earth in relation to the sun has a 41,000-year cycle from 22.1 to 24.5 degrees. This results in more or less sunlight towards the poles.
- The precession, whereby the Earth's tilt "revolves" around its axis on a 26,000-year cycle. This means the North Star will eventually not be the North Star and other stars will take its place in their turn.

The first two of these cycles now appear to have a relationship to the cycles of climate during the Pleistocene.

It was not until the 1970s and through to the mid-1990s that scientists began to take an interest in Milankovitch's discoveries when teams from several countries, including Russia, Japan, the European Union, and Denmark began to drill deep ice cores in the Antarctic and in Greenland. From these ice cores a timeline of various factors – temperature, carbon dioxide and methane in particular – could be determined. The Antarctic cores were drilled more than 3,000 meters (9,840 feet) into the ice sheets, which provided an 800,000-year record of the climate and of CO_2 in Antarctica. It was immediately apparent that the temporal cycles of these and other factors were in close synchronization with the 100,000-year Milankovitch Cycle of orbital eccentricity and that coincided with the periods of maximum glaciation and interglacial periods. Whereas until then there had been no clear explanation for the glacial cycles within the Pleistocene Ice Age, it became clear that they were related to changes in the shape of Earth's orbit around the Sun.

It was eventually determined through marine sediment cores that go back five million years, that for the first 1.6 million years of the Pleistocene Ice Age the glacial cycles were 41,000 years apart and matching perfectly the change in the obliquity (tilt) of the Earth's axis in relation to the Sun's position. It was only then, about one million years ago, that it switched to the 100,000-year cycle of the orbital changes (see Fig. 30, page 66). This is known as the as the "Pleistocene Conundrum" because there is no agreed-upon explanation for it, as is the case for a host of climate-change dynamics.

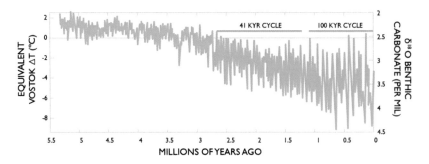

Figure 30. A graph of the Earth's temperature derived from analysis of plankton in ocean sediment cores. The Pleistocene Ice Age began with 41,000-year cycles of glacial maximums, in sync with the change in the Earth's tilt (obliquity) towards the Sun. About one million years ago it shifted to being in sync with the 100,000-year cycles of the shape of the Earth's orbit around the Sun. It is not known why this shift occurred.

For many years climate activists simply took it for granted that the rise in temperature that occurred when a glacial cycle ended was caused by the rise in carbon dioxide that accompanied it. They didn't stop to think that when two factors are strongly correlated it doesn't tell us which one is the cause and which one is the effect, or if there is no cause-effect relationship between them. They just assumed CO_2 was the cause because that was central to the narrative that carbon dioxide is the control knob for temperature. But they were wrong.

This issue was actually at the center of Al Gore's most effective piece of misinformation in his film *An Inconvenient Truth*.[51] He pointed to a graph of CO_2 and temperature that rightly showed they were highly correlated during the glacial cycles, when global temperatures rose and fell by 6-8ºC during each cycle. Al Gore said:

> *The relationship (between CO_2 and temperature) is very complicated, but there is one relationship that is far more important than all the others. When there is more carbon dioxide the temperature gets warmer, because it traps more heat from the Sun inside.*[52]

---

51. Al Gore, *An Inconvenient Truth*, YouTube, May 23, 2012. https://www.youtube.com/watch?v=8ZUoYGAI5iO.
52. Ibid.

He had the audience laughing as he mocked anyone who would question such an obvious conclusion because even a sixth grader could figure this out.

Nowhere in the video does Al Gore mention the Milankovitch cycles even though it was well-known to scientists at the time that the glacial cycles were in sync with the Milankovitch cycles, first with the 41,000-year cycle, then with the 100,000-year cycle. Al Gore must have known about it too, but he probably avoided it because it is one of those "very complicated" aspects of the relationship between carbon dioxide and temperature.

When you think about it, how could small changes in the shape of the Earth's orbit and angle of the tilt cause CO_2 to rise and fall by about 100 ppm in sync with those cycles? However, changes in orbit that alter the distance from the Sun along with changes in tilt that cause solar radiation to strike the Earth at different polar latitudes could plausibly change the temperature regime. And a change in temperature in the atmosphere also means a change in the temperature of the world's oceans causing them to absorb more carbon dioxide when they cool and to emit more carbon dioxide when they warm. Because the oceans contain nearly 50 times as much CO_2 as in the atmosphere, a small change of say one percent in oceanic carbon dioxide causes a nearly 50 percent change in atmospheric carbon dioxide. The 100-ppm change between cycles is just above 50 percent of 180 ppm, and just more than one percent of oceanic CO_2, so the numbers make some sense.

But there is a clincher to this alternative hypothesis, that temperature is the cause and carbon dioxide is the effect during the glacial cycles. A closer analysis of the data from the Antarctic ice cores shows conclusively that the rise in temperature occurs an average of 800 years before the rise in CO_2. One rule in causation is that the effect never comes before the cause, so the fact that carbon dioxide follows temperatures indicates that it is not the cause. The reason there is a relatively long lag time is that when the atmosphere warms it can do so relatively rapidly, but the oceans contain 1,000 times as much heat as the atmosphere and it takes much longer for them to warm or cool. Figure 30 below, showing the ice core data from 50,000 to 2,500 years ago clearly shows that CO_2 follows temperature during the Pleistocene glacial cycles.[53]

---

53. Joanne Nova, "The 800-year lag in CO_2 after temperature – graphed," August 18, 2013. http://joannenova.com.au/global-warming-2/ice-core-graph/.

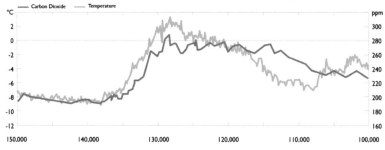

Figure 31. In this graph of Antarctic ice core data from 150,000 to 100,000 years ago, it is clear that temperature leads over carbon dioxide when rising out of the depths of glaciation into the Eemian Interglacial.

To summarize, the 2.6-million-year Pleistocene Ice Age provides ample evidence that atmospheric carbon dioxide concentration during this period has been dictated by the cycles of warming and cooling as the seas emitted and absorbed CO_2 in synchronization with the temperature, with an average lag time of 800 years (see Fig. 31).

The Holocene Interglacial

We are in the Holocene Interglacial today. It began 11,700 years ago after a nearly 10,000-year rise in global temperature coming out of the most recent glacial maximum. The Holocene represents 0.00025 percent of Earth's history and encompasses virtually all of what we consider to be "human civilization." Even though the Holocene is a relatively warm period compared to the longer periods of glaciation during the Pleistocene, it still remains a broadly cold period when compared with most of the history of modern life.

There are a number of approaches to dividing the Holocene into stages. Here we will use the nomenclature adopted in two papers published on Dr. Judith Curry's blog.[54,55] I encourage reading them fully as they have excellent illustrations, a few of which I have included here, and they are very well written.

---

54. Javier, "Nature Unbound III: Holocene climate variability (Part A)," Climate Etc., April 30, 2017. https://judithcurry.com/2017/04/30/nature-unbound-iii-holocene-climate-variability-part-a/.
55. Javier, "Nature Unbound III – Holocene climate variability (Part B)," Climate Etc., May 28, 2017. https://judithcurry.com/2017/05/28/nature-unbound-iii-holocene-climate-variability-part-b/.

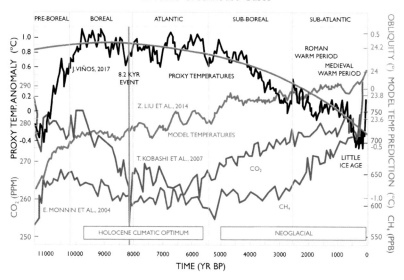

Figure 32. It's a bit busy but there is a lot of information in this chart of key factors during the Holocene. The black line indicates temperature, mainly from ocean sediment cores, as re-analyzed by Javier Viños from data collected by S. A. Marcott et al. The final warming out of the glacial maximum is the Pre-Boreal. The Climatic Optimum comprises the Boreal and Atlantic, and the Neoglacial comprises the Sub-boreal and the Sub-Atlantic. It will come as a surprise to many who have been exposed to the "catastrophic warming" narrative that the climate has been in a net cooling phase for the past 5–6,000 years, and has been in the recent warming period for only the past 300 years, coming out of the Little Ice Age. The red line indicates atmospheric carbon dioxide, which beginning near the close of the Climatic Optimum rose from 260 ppm to 280 ppm while temperature was falling. This is referred to as the Holocene Conundrum because it implies the opposite of the expectations of those who believe higher CO_2 will always result in higher temperature. It does, however, fit the hypothesis that the Milankovitch Cycles are involved as it correlates well with the 41,000-year cycle of the Earth's obliquity (tilt on its axis) which causes a change in the latitude of solar insolation towards the poles (the purple line). The blue line signifies methane (CH_4), also a greenhouse gas that increases along with carbon dioxide and clearly does not cause any warming. Then there is the green line, which illustrates what existing computer models would most likely predict given the programmed-in assumption that increased CO_2 and CH_4 will cause warming. There is no comfort in this 11,500-year record for proponents of the claim that carbon dioxide is causing a climate emergency.*

* Javier, "Nature Unbound III: Holocene climate variability (Part A)," Climate Etc., April 30, 2017. https://judithcurry.com/2017/04/30/nature-unbound-iii-holocene-climate-variability-part-a/.

The Holocene Interglacial is divided into three phases: an initial 2,000-year period of rapid warming, the 4,000-year Holocene Climatic Optimum, followed by the 6,000-year Neoglacial. In addition, there are five vegetative periods based on pollen indicators in sediments: the Pre-Boreal, Boreal, Atlantic, Sub-boreal, and Sub-Atlantic (see Fig. 32).

It is typical for at least the past few interglacials for the temperature to rise out of the glacial maximum for about 10,000 years and to reach the interglacial period's maximum temperature early in that period. During the Climatic Optimum the average temperature of the Earth

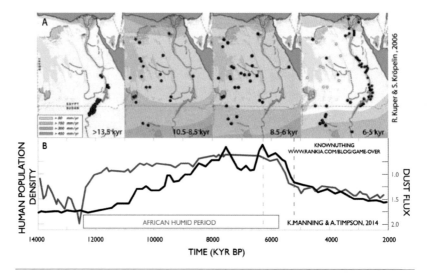

Figure 33. A sequence of maps of the Eastern Sahara showing the period of maximum greening and population during the Climatic Optimum between 14,000 and 2,000 years ago. Red dots are settlements based largely on livestock herding. As the Neoglacial set in, the Sahara reverted to a desert, as it had been throughout the previous glacial period. The population then concentrated in the Nile River Valley; concurrent with the rise of the Egyptian Empire. The graph herein shows human population density caused by human settlement in black, and dust from livestock in red.*

* Ibid.

was at least 1°C (1.8°F) warmer than today. And at least on a regional level, it was wetter than the most recent 6,000-year Neoglacial as the Sahara Desert and other regions were green. There were towns and livestock-herders across the Sahara's vast expanse (see Fig. 33). Hence the term "Climatic Optimum," as it was the most pleasant climate since the previous interglacial period 110,000 years earlier.

The Neoglacial (literally new glacial) phase of the Holocene marks a break in the relatively stable warmer climate of the Climatic Optimum. The Neoglacial is characterized by a gradual descent into the coldest period since the early beginnings of the Holocene. The Little Ice Age, which reached its coldest around 1650-1700 AD, followed the Medieval Warm Period, when the Vikings colonized and farmed southern Greenland. We are presently in the Modern Warm Period, but the climate is not as warm today as it was during the Climatic Optimum (see Fig. 34). If the pattern of the past 6,000 years repeats itself the climate will again turn to cooling in about 150 years.

The Modern Warm Period began when the Little Ice age peaked around 1700, and the climate began to warm again. Human emissions of carbon dioxide from 1700 to 1850 were insignificant and yet historical records indicate the Earth warmed at about the same rate during

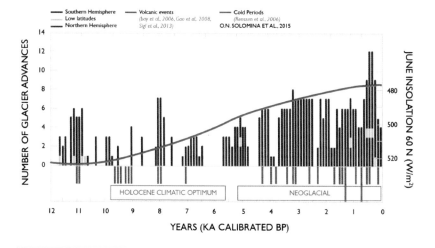

Figure 34. This chart shows the number of glacial advances per century within the 17 main global glacial zones during the past 12,000 years. The red bars indicate the 12 northern hemisphere zones, the blue bars indicate the four southern hemisphere zones, and the yellow bars indicate the one low-latitude zone. The gray line shows the change in solar insolation at 60° north latitude caused by the change in the obliquity (tilt) of the Earth's axis. The correlation between the tilt and the glacial advances is striking.*

* Javier, "Nature Unbound III: Holocene climate variability (Part A)," Climate Etc., April 30, 2017. https://judithcurry.com/2017/05/28/nature-unbound-iii-holocene-climate-variability-part-b/

that period as it has since; from 1850 to the present (see Fig. 35, page 72). Regardless, the "CO_2 is the control knob of Earth's temperature" advocates are inferring that a very short-term correlation implies a very long-term causation when this is simply not the case.

The interglacial period preceding the Holocene, called the Eemian, lasted for 15 thousand years from 130 to 115 thousand years ago. During the Eemian, the sea level was between five to eight meters (16.5 to 26 feet) higher than it is today.[56] The global temperature was between 2-4°C (3.6-7.2°F) higher than it is today, even though at that time carbon dioxide was only at 280 ppm, about the same as it was in pre-industrial 1750 when it was even colder than today. This one example discredits the idea that CO_2 is the leading cause for the Earth's current warming.[57] The two interglacials that preceded the Eemian, the La Bouchet and the Purfleet, were also warmer and had higher sea levels than the Holocene. In 2020, with carbon dioxide at 415 ppm, the

---

56. R. Kopp, et al., (2009). "Probabilistic assessment of sea level during the last interglacial stage." *Nature*, 462: pp863-86. https://www.nature.com/articles/nature08686.
57. National Aeronautics and Space Administration – Goddard Institute for Space Studies, "How Carbon Dioxide Controls Earth's Temperature," October 14, 2010. https://www.giss.nasa.gov/research/news/20101014/.

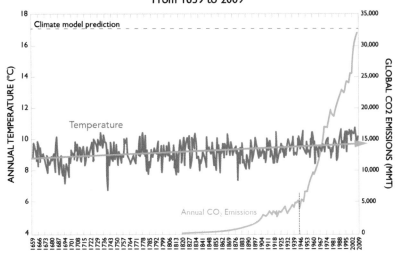

Figure 35. This is the longest known continuous measurement of temperature using a thermometer, from Central England, showing the continuous slow rise of temperature during the past 300 years. The exponential rise in carbon dioxide emissions beginning in 1850 has not been matched by a similar response from temperature. The spurt in temperature rise from 1694 to 1729 was longer and more pronounced than any rise since 1900. This gives no support for the CO_2-caused climate catastrophe storyline. The CO_2 Coalition has published a statistical analysis for the years 1900 to 2018 that clearly demonstrates that CO_2 is not the cause of the temperature increase during this period.*
* Caleb Rossiter, PhD, "Equal Warming, 1900 to 1950 versus 1950 to 2018: Why the UN Knows the First Half was Natural," CO_2 Coalition, April 9, 2020. http://co2coalition.org/publications/equal-warming-1900-to-1950-versus-1950-to-2018-why-the-un-knows-the-first-half-was-natural/.

Holocene is still much cooler, and the sea level much lower than during these three preceding interglacials when CO_2 was at 280-300 ppm (see Fig. 36). The alarmists have no answer to this, they simply ignore it, as they do so many other telling facts about Earth's climate before the industrial era.

The International Commission on Stratigraphy

Stratigraphy - def. 'the branch of geology concerned with the order and relative position of strata and their relationship to the geological time scale.'[58]

---

58. "Definition of stratigraphy in English," LEXICO, 2020. https://www.lexico.com/en/definition/stratigraphy.

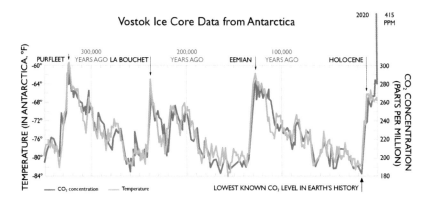

Figure 36. A graph of temperature and carbon dioxide data from the Vostok Ice Cores in Antarctica shows the four most recent 100,000-year cycles of glaciation and interglacial climates. All three of the interglacials prior to the present Holocene were warmer and had considerably higher sea levels despite the fact that CO_2 levels were comparable with the pre-industrial Holocene. Even though carbon dioxide has risen to 415 ppm by 2020 there is no indication that temperature is following the CO_2 upward in such a dramatic fashion.

The International Commission on Stratigraphy (ICS) is the largest and oldest constituent scientific body in the International Union of Geological Sciences. It is responsible for identifying and naming the many periods of geological and evolutionary change through the study of rock layers, sediments, and the fossils within those sediments. This gives them a responsibility that covers the entire 4.6-billion-year history of the Earth. They have done an excellent job over the years and their archives contain a massive trove of factual information about the vast number of events from volcanoes to asteroids to ice ages to the full evolution of life. But even these scholars have not been immune to the tactics of the climate alarmists who insist we are about to change the climate like it has never changed before, and that this calls for drastic changes to the way we name the ages of the Earth. They are convinced we are heading for certain doom if we don't end the use of fossil fuels, demanding a "cure" that would certainly be far worse for civilization than the actual alleged "disease."

As can be seen from Figures 37 and 38 below, the Pleistocene Ice Age, that we have been in for the past 2.6 million years, has now been declared to have ended and is followed by the Holocene Epoch; which until recently, was only considered one of the interglacial periods – along with the 44 or so other interglacial periods – in the Pleistocene Epoch. In climatic terms, the Holocene is in no way different from the interglacial periods before it. In fact, as shown in the previous section, the Holocene has not been as warm as the previous three interglacials, even during the Holocene Climatic Optimum when the Earth was actually warmer than it is today. And although the Holocene has experienced a significant cooling trend during the past 6,000 years, the ICS clearly succumbed to pressure and declared the current interglacial an Epoch when previously none of the numerous interglacial periods had any official designation. They were simply part of the Pleistocene Epoch, usually known as the Pleistocene Ice Age because that is what it is. None of the other interglacials have been designated as epochs and only the past few, prior to the Holocene, have even been assigned names.

It appears that the primary rationale for declaring the Holocene Interglacial an epoch is the advent of human civilization and the belief that we will alter the climate so dramatically that another glacial maximum is virtually impossible. Apparently the "Age of the Humans" has ushered in a complete alteration of global geology, global climate,

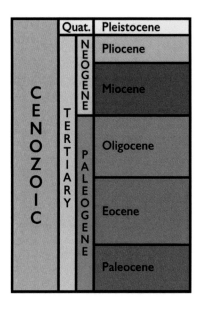

Figure 37. This is what the International Commission on Stratigraphy geological timeline for the Cenozoic Era looked like before it was decided they would rename the Holocene Interglacial as an Epoch. The Holocene was previously treated as part of the Pleistocene Epoch, like all the other 44 or so interglacial periods that preceded it during the Pleistocene Ice Age. There appears to be no record on the Commission's website of when the Holocene was declared an Epoch. A concerted search of the Internet provided no information about when this change was made or why it was made. Source: https://ucmp.berkeley.edu/cenozoic/cenostrat.html.

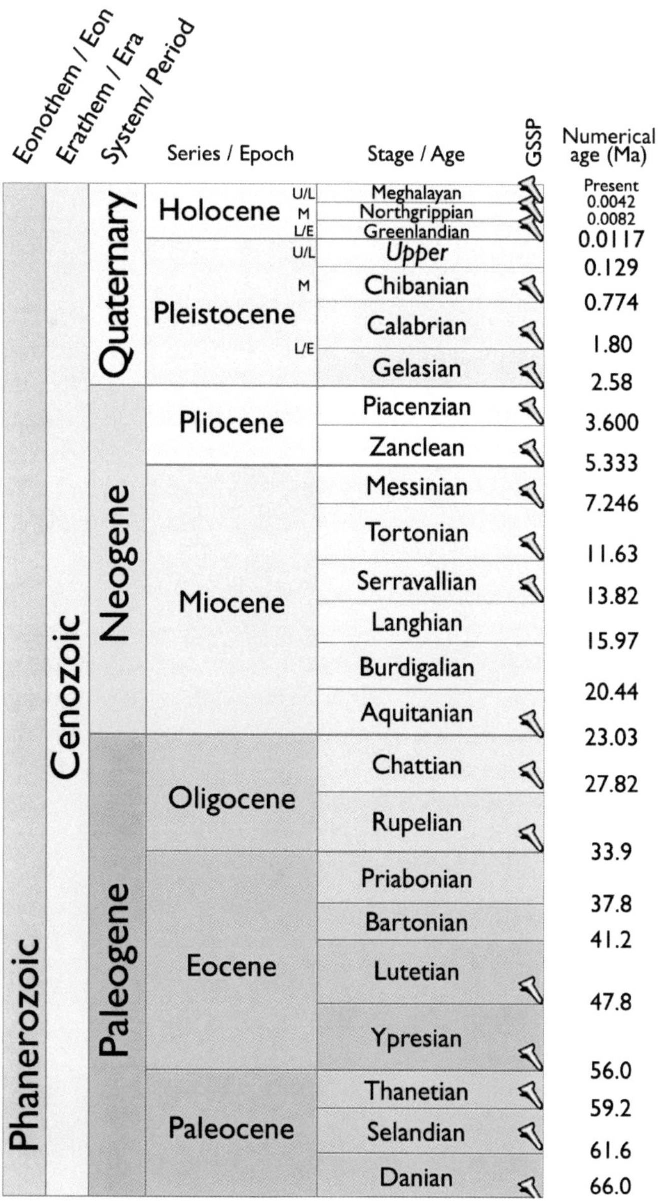

Eonothem / Eon	Erathem / Era	System / Period	Series / Epoch		Stage / Age	GSSP	Numerical age (Ma)
Phanerozoic	Cenozoic	Quaternary	Holocene	U/L	Meghalayan		Present 0.0042
				M	Northgrippian		0.0082
				L/E	Greenlandian		0.0117
			Pleistocene	U/L	Upper		0.129
				M	Chibanian		0.774
				L/E	Calabrian		1.80
					Gelasian		2.58
		Neogene	Pliocene		Piacenzian		3.600
					Zanclean		5.333
			Miocene		Messinian		7.246
					Tortonian		11.63
					Serravallian		13.82
					Langhian		15.97
					Burdigalian		20.44
					Aquitanian		23.03
		Paleogene	Oligocene		Chattian		27.82
					Rupelian		33.9
			Eocene		Priabonian		37.8
					Bartonian		41.2
					Lutetian		47.8
					Ypresian		56.0
			Paleocene		Thanetian		59.2
					Selandian		61.6
					Danian		66.0

Figure 38. This is the depiction of the Cenozoic Era on the Commission's website today. This has the effect of declaring the Pleistocene Ice Age as having ended. There is no basis in fact to make such a judgement. The Holocene is in no way fundamentally different from the other 44 or so interglacial periods that have occurred in the Pleistocene.

Note that the Pleistocene and the Holocene are now divided into Ages, which did not exist before. These Ages are much shorter in time than any previous Ages, most of which lasted for millions or tens of millions of years. The previous 44 or so interglacial periods are not identified within the Pleistocene. Source: https://stratigraphy.org/icschart/ChronostratChart2020-01.jpg.

and is triggering the "sixth mass extinction"[59] (this will be the subject of another chapter).

But we have not actually altered the climate in any way out of the ordinary and there is no hard evidence that we will. The climate of Earth today is not at all unusual for an interglacial period. Therefore, the designation of the Holocene as an epoch is scientifically dishonest and was obviously done for political and financial reasons rather than scientific ones. The fact that the decision appears to be buried from view and there is no explanation for the change reinforces this conclusion. An internet search for "when did the International Commission on Stratigraphy declare the Holocene to be an epoch?" produces no results, never mind finding the reasons for them making the change. We might appreciate an answer to these questions. I have emailed them with this question and received no reply.

But it gets worse. In recent years there has been a strong lobbying effort by the climate alarmist community to declare an end to the Holocene Interglacial "Epoch" and declare the advent of a brand-new epoch, the "Anthropocene," which is literally translated to the Age of the Humans. One suggestion is that the new epoch be declared to have begun in 1950, when radioactive elements from nuclear-weapons testing could first be observed in ocean and land sediments. A sub-committee of the ICS formally recommended the adoptions of the Anthropocene to the board of the ICS in 2018. Thankfully they declined, instead dividing the Holocene into three new ages; the Greenlandian Age, the Northgrippian Age, and the Meghalayan Age. The proponents for the Anthropocene were aghast at what they considered to be an insult to their noble cause. The Global Warming Policy Forum in London, England, quoted senior editor Graham Lloyd of the *Australian*, who did an excellent job of documenting this tempest in a teapot.[60]

In the end it comes down to the little slip made by researchers at the University of California Berkeley. They stated, "but in general, the Holocene has been a relatively warm period in between ice ages," implying that another glacial period (ice age) would follow the Holocene. There is no doubt that the Holocene should have been left with its original status as one of many interglacial periods until, and unless, it became abundantly clear that the climate had changed so drastically

---

59. Gerardo Ceballos, et al., "Biological annihilation via the ongoing sixth mass extinction signaled by vertebrate population losses and declines," Proceedings of the National Academy of Sciences, July 2017, p114 (30) E6089-E6096; DOI: 10.1073/pnas.1704949114. https://www.pnas.org/content/114/30/E6089.
60. Global Warming Policy Forum, "Is It All Over For The 'Anthropocene' Promoters?" August 12, 2018. https://www.thegwpf.com/is-it-all-over-for-the-anthropocene-campaigners/.

that it was, in fact, coming out of the Pleistocene for good. That would require an increase in global temperature of at least 5°C (9°F), not the 1.2°C (2.2°F) that has occurred during the 300 years of the Modern Warm Period, none of which can be proven to be caused by our CO_2 emissions. Such a rise would still be well within long-term historical norms. The Eocene Thermal Maximum, which the ancestors of every species on Earth today lived through, was as much as 10°C (18°F) higher than today's global average. For the time being, there is no evidence that the Pleistocene has ended, and that is all there is to it.

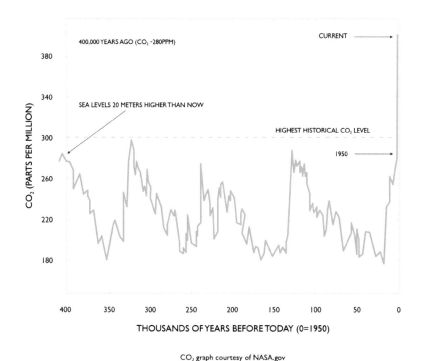

Figure 39. This graph, adapted by NASA, illustrates two important points. First, about 400,000 years ago during an interglacial similar to the one we are in now, the sea level was 20 meters (66 feet) higher than today even though atmospheric carbon dioxide was only at 280 ppm – about the same as pre-industrial levels during this interglacial. Second, carbon dioxide has shot up to more than 400 ppm during the past 150 years and this has not resulted in any change in the current slow rate of sea level rise, nor has global temperature followed the steep rise in CO_2. These two factors alone should incite speculation about the certainty of the claim that carbon dioxide is the cause of both the slight global temperature rise and the modest rate of sea level rise.

The Impact of the Fake Climate Catastrophe Campaign on Society

Whereas more CO_2 is entirely beneficial for life including the plants we depend on, the biggest threats to society and the environment, are the very policies that are being adopted to "fight catastrophic climate change." Near the top of the list is the widespread adoption of "renewable energy" – in particular wind and solar – devices which have nothing renewable in their machinery. These two unreliable technologies have caused the price of electricity to double, and more in some countries.

There is a built-in energy cost every time we make something or move something. There is a large energy cost to simply staying alive in all climates outside the tropics. Energy underlies everything about society and our economy and unless we want to go back to subsistence agriculture where 75 percent or more of the population is engaged in the back-breaking work of growing food we must either continue with the large-scale use of fossil fuels or multiply the number of large nuclear plants by about 20 times the present 440 worldwide. More on that in another chapter.

Nearly all the decisions being made to stave off the "climate catastrophe" are having far more negative impacts on poor people than on wealthy people. Poor People Matter, and it has nothing to do with their skin pigment. They are the first ones to have their power cut off when they decide to prioritize buying food for their family. This is particularly true for African countries where policies adopted by international agencies, such as the World Bank, forbid investment in clean-burning fossil-fuel plants. Except for hydroelectric power, which is the most economical on good sites, fossil fuels are the most economical source of reliable electricity in most regions.[61]

The World Bank stated:

The World Bank is committed to helping countries implement economically smart and tailored approaches that best suit their needs, and supports technological, financial, and policy innovations that can help accelerate the expansion of reliable and affordable electricity services and end energy poverty.

---

61. The World Bank, "This Is What It's All About: Boosting Renewable Energy in Africa," February 26, 2019. https://www.worldbank.org/en/news/feature/2019/02/26/this-is-what-its-all-about-boosting-renewable-energy-in-africa.

However, on the contrary they also stated:

> *The World Bank Group has not financed a new coal-fired power plant since 2010 and has no active coal-fired power generation in its pipeline. The Bank will support countries transitioning away from coal by helping close coal mines and ensure a just transition for affected communities.*[62]

In other words, the World Bank intends to make electricity much more expensive than it is now for a region that has more poor people per capita than any other region in the world.

---

62. The World Bank, "Energy," July 13, 2020. https://www.worldbank.org/en/topic/energy/overview#2.

CHAPTER 4

Polar Bears are Threatened with Extinction Because of Climate Change

Here's a doomsday story that's been relentlessly drilled into the collective consciousness of adults and children alike. It's a real candidate for the most fake news story of all. There is absolutely no truth to the claim that polar bears are endangered by anything, including climate change.

As a matter of fact, the only reason polar bears exist at all is precisely because of climate change. The European brown bear (called the grizzly bear in North America) was one of the species of wildlife that came to the New World, along with humans, over the Bering Land Bridge from Asia about 15,000 years ago. But some time before that, during one of the glacial periods of the Pleistocene Ice Age, the polar bear (Ursus maritimus) evolved from the brown bear as a distinct species with white fur which has an advantage as camouflage in snow and ice. Their metabolism changed to support a largely carnivorous diet as there were few plants or berries to eat in the Arctic. Even after hundreds of thousands of years of geographical separation and very different dietary habits, grizzly bears occasionally breed successfully with polar bears and produce viable offspring, although it doesn't happen very often.[1]

---

1. University of Buffalo, "Polar bear evolution tracked climate change," *Science Daily*, July 23, 2012. https://www.sciencedaily.com/releases/2012/07/120723151028.htm.

Figure 40. Unlike most bear species, polar bears are at home in the ocean and are nicknamed "Sea-Bears." The idea that they are now drowning due to shrinking ice floes is certainly fake news.* The ice has receded every summer for millennia. Of course, it is possible that polar bears might drown occasionally, just like humans do. Veritably, 1.2 million people drown each year and surely that is not blamed on climate change. Whoops, actually the World Health Organization predicts more drowning from flooding caused by climate change.**

* David Derbyshire, "Scientist who claimed polar bears were drowning is investigated for 'scientific misconduct,'" Daily Mail, July 28, 2011. https://www.dailymail.co.uk/sciencetech/article-2019953/Scientist-claimed-polar-bears-drowning-investigated-scientific-misconduct.html
** World Health Organization, "Climate change and health," February 1, 2018. https://www.who.int/news-room/fact-sheets/detail/climate-change-and-health

Polar bears would not exist today if it were not for the profound cooling of the climate which began 50 million years ago and set in dramatically about five million years ago in the Northern Hemisphere. Many of the plants and animals that live in the Arctic today evolved and adjusted to the new climate, or moved to a warmer climate, as species have done throughout the history of life. You would never know it listening to the media today, but polar bears have experienced a dramatic increase in their population in recent decades all around the Arctic. Let's begin with a repeat of a graph from the chapter on climate change (see Fig. 41, page 82).

That the Earth is at the tail-end of a 50-million-year cooling period is not disputed, it is simply ignored and substituted with the comparatively minuscule increase in temperature during the past 300 years, which is not very relevant in terms of evolutionary adaptation. But there is a relatively recent event that the public has been shielded from since the polar bear "extinction" myth was invented about 20 years ago.

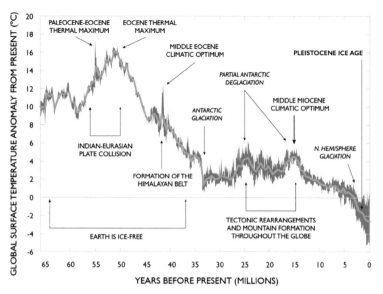

Dynamics of global surface temperature during the Cenozoic Era reconstructed from ¹⁸O proxies in marine sediments (Hansen et al. 2008)

Figure 41. A graph of global temperature during the past 65 million years, showing the Eocene Thermal Maximum when the land near both poles was ice-free and forested and the oceans were much warmer. The Eurasian brown bear (grizzly bear) was a far northern species where glaciation did not set in until about three to four million years ago. It was not until sometime after that point that the polar bear evolved from the brown bear. A number of genetic studies have tried to nail down the timing of this transition but so far all have come up with a wide range of estimates, ranging from about four to five million years ago to 134 thousand years ago, although somewhere between 600 thousand and 160 thousand are the most plausible* (After Hansen, et al.)**

* Frank Hailer, et al., "Nuclear genomic sequences reveal that polar bears are an old and distinct bear lineage," PubMed, April 20, 2012. https://pubmed.ncbi.nlm.nih.gov/22517859/olismge.

** Hansen, et al., "Target Atmospheric CO_2: Where Should Humanity Aim?, *The Open Atmosphere Science Journal*, 2008, 2, pp217-231. https://pubs.giss.nasa.gov/docs/2008/2008_Hansen_ha00410c.pdf.

When I speak at conferences of 100 to 1,000 highly educated science and industry professionals I have often asked, "How many of you know about the international treaty signed in 1973 among all polar countries in which they outlawed the unrestricted hunting of polar bears?" It is unusual for even one person to raise their hand. That is because the media rarely, if ever, mentions it. They parrot the doomsday predictions of "scientists" who are on serial government grants traveling to the Arctic for a few weeks each year to "study" the bears.

This treaty was negotiated and signed into international law in Oslo, Norway on November 15, 1973 by the governments of Canada, Denmark, Norway, the Union of Soviet Socialist Republics, and the

United States of America.² This includes all countries with polar bear populations, Denmark's being in Greenland. The treaty was a response to the increase in the harvest of polar bears by recreational hunters and signs that the population was declining rapidly. Some members banned polar bear hunting altogether, like Norway, while some, like Canada, set a strict limit to hunting and required an Inuit guide to accompany the hunters.

Contrary to the alarmist rhetoric of the past two decades this has proved to be one of the most successful conservation initiatives during the past century. In 1973 the circumpolar population of polar bears was estimated to be as low as 6,000 to about 12,000. Today in 2020, the official estimate is between 22,000 to 31,000 but may be much higher; perhaps four to five times as many bears as 47 years ago.³ The fact is, environmental activists, politicians, the media, and many scientists have purposefully engaged in a misinformation campaign for their own personal financial and political interests rather than telling the truth. They have fabricated a fake catastrophe that the average citizen could not validate through independent observation. It is impossible to see if carbon dioxide is the cause of melting ice, it is difficult to know if melting ice is a threat to the bears, and it is likewise impossible for the public to count all the polar bears around the north pole.

Let's look at the widely distributed claim that melting sea ice threatens polar bears with extinction. *National Geographic* has led the way, using images of a starving polar bear and claiming, "This is what climate change looks like" (see Fig. 42, page 84).⁴ When they were challenged, the editors – nearly nine months later – admitted they did not know why this particular bear was starving and still reverted to the claim that polar bears in general were starving to death due to melting ice caused by climate change.⁵ This completely contradicts ample first-hand evidence that polar bears today are generally fat and healthy, and that the *National Geographic* photographer found only one bear that was starving. Of course, every polar bear gets old if they are lucky,

---

2. Polar Bear Range States, "The 1973 Agreement on the Conservation of Polar Bears," June 20, 2017. https://polarbearagreement.org/resources/agreement/the-1973-agreement-on-the-conservation-of-polar-bears.
3. Susan J. Crockford, "State of Polar Bear Report 2019," Global Warming Policy Forum, 2019, ISBN978-0-9931190-7-1. https://polarbearscience.files.wordpress.com/2020/02/crockford-2020_statepb2019-final.pdf.
4. Jesse Ferraras, "This polar bear is starving, but it's not 'what climate change looks like': National Geographic," Global News, August 1, 2018. https://globalnews.ca/news/4361868/polar-bear-climate-change-national-geographic/.
5. Steven Leahy, "Polar Bears Really Are Starving Because of Global Warming, Study Shows," *National Geographic*, February 2018. https://www.nationalgeographic.com/news/2018/02/polar-bears-starve-melting-sea-ice-global-warming-study-beaufort-sea-environment/.

Figure 42. A clip from the video of the starving polar bear featured in the December 2017 issue of *National Geographic*. It took nearly nine months for *National Geographic* to admit there was no basis for that claim that this was caused by climate change. The video has been viewed on YouTube by 2.3 million people. The video was still up as of November 3, 2020.*

* "Heart-Wrenching Video: Starving Polar Bear on Iceless Land," *National Geographic*, December 11, 2017. https://www.nationalgeographic.co.uk/animals/2018/02/polar-bears-really-are-starving-because-global-warming-study-shows.

and most eventually die of starvation. It is the leading natural cause of polar bear death.[6] There are no nursing homes for wild, old polar bears. Apparently, the editors of *National Geographic* don't know this.

There is an assumption made and constantly reinforced, that the more sea ice there is in the Arctic the more seals there will be for the polar bears to eat. This reflects a failure to understand the ecology of the Arctic. Where there is sea ice, especially multi-year ice which is quite thick, not much sunlight gets to the ocean below to support the growth of phytoplankton, which are the entire basis of the food chain in the Arctic Sea. While some types of phytoplankton can grow under the ice, they are often limited in abundance. Only when a considerable area of the sea is open water, especially in the summer months, is there maximum growth of the plankton. If there is less phytoplankton

6. S. C. Amstrup, "Polar bear (Ursus maritimus)" *Wild Mammals of North America*, G. A. Feldhamer, et al., (eds), pp587-610, Johns Hopkins University Press, Baltimore, January 1, 2010. www.fws.gov/ecological-services/es-library/pdfs/Polar-Bear-SBS-Final-SAR.pdf.

there will be less zooplankton which feed on them, and if there is less zooplankton there will be fewer fish. If there are fewer fish there will be fewer seals, and if there are fewer seals there will be fewer polar bears. That's how the food chain works.

Therefore, it is not the case that more sea ice automatically means there will be more seals and polar bears. There is clearly a range of conditions somewhere between "all ice" and "no ice" that provide the optimum support for seal production, and seeing the polar bear population has grown to four or five times what it was when the treaty was signed in 1973, there is no basis for alarm, and certainly no basis for predictions of extinction. Here is a passage from Susan Crockford's paper, "State of Polar Bear Report 2019":

> *Between 2007 and 2015, summer sea ice on average dropped about 38 percent from 1979 levels, an abrupt decline to within measurement error of the reduced coverage expected to occur by mid-century. Christine Hunter and colleagues proclaimed in 2007 that such reduced summer sea ice by 2050, if present for eight out of ten years (or four out of five years), would generate a massive drop in polar bear numbers: ten vulnerable subpopulations out of 19 would be extirpated* (locally extinct), *leaving fewer than 10,000 animals worldwide (a 67 percent decline). Even though summer sea ice from 2016 to 2019 has continued this pattern, recent research shows such a decline in polar bear abundance has not occurred. This indicates summer sea ice levels are not as critical to polar bear survival as USGS biologists assumed. Despite marked declines in summer sea ice, Chukchi Sea polar bears continue to thrive, and reports from a survey of Wrangel Island bears in the fall of 2019 showed bears were abundant, healthy, and reproducing well, as bears in the US portion were in 2016. Similarly, according to Jon Aars, a senior Norwegian biologist, polar bears in the Svalbard area show no impact of the particularly low sea ice years of 2016 to 2018, and 2019 has proven no different.*[7]

This soundly corroborates that extensive summer sea ice may, in fact, be a detriment to maximum productivity in the sea and that a minimum of sea ice in the summer may result in higher productivity so long as there is sufficient sea ice during the winter and spring months, which is when polar bears return to the ice to hunt seals.

The predictions of an "ice-free" Arctic are nearly always couched with "in the summer months," as a return to an ice-free Arctic

---

7. Susan J. Crockford, "State of Polar Bear Report 2019," Global Warming Policy Forum, 2019, ISBN978-0-9931190-7-1. https://polarbearscience.files.wordpress.com/2020/02/crockford-2020_statepb2019-final.pdf.

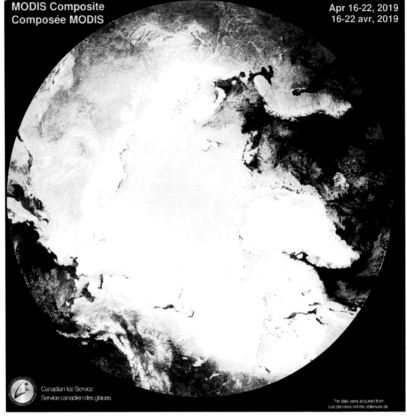

Figure 43. This shows the extent of ice and snow in the Arctic and beyond between April 16th and the 22th in 2019. There is no indication that winter ice will disappear in the Arctic in the foreseeable future. Therefore, predictions of a drastic decline or the extinction of polar bears are entirely without foundation.

year-round would require an end to the Pleistocene Ice Age and a return to the climate of about five to ten million years ago. Not only has ice persisted in the summer months, the Arctic has remained heavily covered in ice during the winter and spring (see Fig. 43).

Until quite recently, despite much evidence to the contrary, many polar bear scientists continued to predict a decline in the population. Thanks largely to the efforts of Dr. Susan Crockford of Victoria, British Columbia, these scientists have been obliged to accept the fact that polar bear populations are robust and are continuing to grow in numbers as indicated by the most recent census (see Fig. 44).

But Dr. Crockford's determination over the years to see the truth told was not without repercussions. In April 2017, the journal *Bioscience*

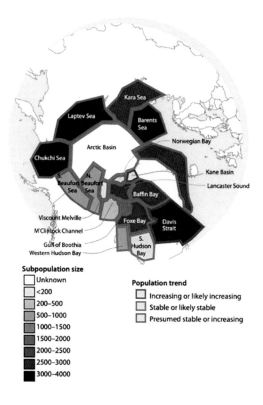

Figure 44. Finally, the scientists who have been predicting the early demise of polar bears have recognized that they are not endangered. Dr. Susan Crockford can take the credit for this for tirelessly speaking and writing the truth. There is absolutely no reason to believe the continued claims that, despite their increase in population since the treaty to stop unrestricted hunting of polar bears was signed in 1973, they are still endangered because of climate change.

published a paper titled "Internet Blogs, Polar Bears, and Climate-Change Denial by Proxy" written by Jeffrey A. Harvey and 13 others.[8] Two of these authors are members of the Polar Bear Specialists Group, which operates by "consensus," meaning you either agree with the group or you get the boot. One of the others was Michael E. Mann who is known for his vicious attacks on any scientist who does not bow to the narrative of climate catastrophism. He is decidedly not an expert on polar bears. Dr. Crockford's name was mentioned 19 times in the article, and she was labeled a "science-denier." The word "denier" appeared 31 times in the paper, most often with reference to Dr. Crockford. The

---

8. Jeffrey A. Harvey, et al., "Internet Blogs, Polar Bears, and Climate-Change Denial by Proxy," *Bioscience*, Volume 68, Issue 4, April 2018, pp281-287 (originally published November 29, 2017. This is a "corrected" version that attempts to discredit Dr. Crockford even further). https://doi.org/10.1093/biosci/bix133.

paper divided the numerous "blogs" (personal websites) about climate change and polar bears into "science-based blogs" and "denier blogs." The reader will not find it hard to guess which category Dr. Crockford's blog fell into. All-in-all it was a vicious smear job with all the usual suspects piling on. I highly recommend Dr. Crockford's website.[9]

It is very unusual that a respectable journal would publish what is so obviously a blatant political attack, using the word "denier" without being clear about exactly what that term means. One can infer that it means she does not agree with their interpretation of the data around polar bear populations and summer sea ice cover. At this time, all but a few polar bear populations studied are stable, likely increasing, or actually increasing. It is known that summer sea ice was markedly reduced from 2007 to 2019, which happened decades before it was expected. Sea ice experts had predicted it wouldn't happen until 2050. US Geological Survey biologists predicted a massive decline in polar bear numbers in 2007, but it did not occur.[10] Dr. Crockford was correct in her conclusion that their prediction was wrong.[11] Then what is this mob accusing her of "denying"?

In their 2018 "corrections" the authors stated that:

Crockford has neither conducted any original research nor published any articles in the peer-reviewed literature on the effects of sea ice on the population dynamics of polar bears.[12]

This is Dr. Crockford's reply to this claim in an interview with Anthony Watts. The entire interview is well worth a read as an example of the cancel culture prevalent among climate alarmists and political extremists today.[13] Here is an excerpt from it:

The dissertation I wrote for my PhD on speciation included a discussion of polar bears (Crockford 2004).[14] In addition, I have an article on evolution

---

9. S. J. Crockford, "*Polar Bear Science – Past and Present,*" Internet Blog. https://polarbearscience.com/
10. S. C. Amstrup, et al., "Forecasting the rangewide status of polar bears at selected times in the 21st century," US Geological Survey. Reston, VA.; Courtland, *Nature 453*, pp432-433, May 22, 2007. www.fs.fed.us/pnw/pubs/journals/pnw_2007_amstrup001.pdf.
11. S. J. Crockford, "The Polar Bear Catastrophe That Never Happened," Global Warming Policy Foundation, 2019. https://www.amazon.com/Polar-Bear-Catastrophe-Never-Happened-ebook/dp/B07PT7SCZ8.
12. Jeffrey A. Harvey, et al., "Corrigendum: Internet Blogs, Polar Bears, and Climate-Change Denial by Proxy," *BioScience*, Volume 68, Issue 4, April 2018, p237. https://doi.org/10.1093/biosci/biy033.
13. Anthony Watts, "An interview with Dr. Susan Crockford on the Harvey, et al., attack paper over polar bear research," December 7, 2017. https://wattsupwiththat.com/2017/12/07/an-interview-with-dr-susan-crockford-on-the-harvey-et-al-attack-paper/.
14. S. J. Crockford, "Animal Domestication and Vertebrate Speciation: A Paradigm for the Origin

in a peer-reviewed journal in which polar bears are prominently featured (Crockford 2003),[15] and two official comments, with references, on polar bear hybridization (which is how these were handled in these two journals at the time (although some have argued these are not strictly peer reviewed, they were vetted by the journals at the time: it wasn't like posting a comment on a blog, they had to be approved). I also have a paper in a peer-reviewed book chapter on ringed seals, the primary prey of polar bears (Crockford and Frederick 2011),[16] and a peer-reviewed journal article on the paleohistory of Bering Sea ice, the habitat of Chukchi Sea polar bears (Crockford and Frederick 2007).[17]

While it is true that these peer-reviewed papers are not the result of field or laboratory research on polar bears and most do not focus exclusively on polar bears, they do deal with the history of polar bear habitat, the ecology and physiology of their primary prey, and the evolution of polar bears as a species (which requires a firm understanding of their zoogeography, ecology, genetics, physiology, behavior, and life history). I don't believe that the definition of a peer-reviewed paper on polar bears implies it be only about polar bears. These topics are all valid aspects of polar bear biology and cannot be dismissed as irrelevant to my expertise.[18]

Unfortunately for Dr. Crockford, the nastiest cut was yet to come. First, after many years of speaking engagements for which she was not paid, she was excluded from the University of Victoria's speaker's bureau when she was accused by outsiders of "bias." Then in October 2019, the same university refused to renew her appointment as an adjunct professor, a position she had held for 15 years. No reason was given, and

---

of Species," 1976. https://dspace.library.uvic.ca/bitstream/handle/1828/542/crockford_2004.pdf?sequence=1&isAllowed=y.

15. S. J. Crockford, "Thyroid hormone phenotypes and hominid evolution: a new paradigm implicates pulsatile thyroid hormone secretion in speciation and adaptation changes," *International Journal of Comparative Biochemistry and Physiology* Part A 135(1): pp105-129, 2003. http://europepmc.org/article/med/12727549.

16. S. J. Crockford and G. Frederick, "Neoglacial sea ice and life history flexibility in ringed and fur seals," T. Braje and R. Torrey, eds. "Human and Marine Ecosystems: Archaeology and Historical Ecology of Northeastern Pacific Seals, Sea Lions, and Sea Otters," U. California Press, LA, pp65-91, 2011. https://www.ucpress.edu/book/9780520267268/human-impacts-on-seals-sea-lions-and-sea-otters.

17. Susan Crockford and S. G. Frederick, "Sea ice expansion in the Bering Sea during the Neoglacial: Evidence from archaeozoology," Researchgate, August 2007. https://www.researchgate.net/publication/249664555_Sea_ice_expansion_in_the_Bering_Sea_during_the_Neoglacial_Evidence_from_archaeozoology.

18. It is important to note that with the awarding of a PhD, the document itself does not provide the specific area of research or study involved. For example, there is no such thing as a "PhD in polar bears" even if that was the subject of one's PhD thesis (dissertation). PhD stands for Doctor of Philosophy, or in plain English "lover of wisdom." It denotes more than a narrow academic discipline, rather it indicates the ability to research, interpret, and understand a wide range of subjects in one's quest for knowledge.

the university had the nerve to proclaim, "the University of Victoria, in both word and deed, supports academic freedom and free debate on academic issues."[19] Because she now has no university affiliation, Dr. Crockford has lost her ability to apply for research grants, her use of the university library, and her ability to collaborate with other university-affiliated scientists. She describes her treatment by the university as "an academic hanging without a trial, conducted behind closed doors."[20]

There are about 150,000 Inuit people living in and near the Arctic; about 65,000 in Canada, 50,000 in Greenland, and about 16,000 in each Alaska and Russia.[21] Their history goes back three to four thousand years, so they have seen a polar bear or two. Today, most of these people spend the entire year in small coastal villages where they experience the full cycle of life and depend upon it for much of their food. Before European contact they depended on polar bears for meat, fur, and bones. Their sled dogs would surround the bears and harass them so they could be speared, or they hunted females in their winter dens. While these methods are no longer practiced, the Inuit maintain a strong interest in polar bears, especially when they kill members of their villages.

The northeast territory of Canada is named Nunavut where there are no trees due to the cold climate, and this is where most of the Canadian Inuit live. It includes Baffin Island where the capital Iqaluit is situated. In November 2018, the government of Nunavut tabled a draft management plan for polar bears which made the case that there are now so many bears that they have become a safety hazard for the villagers (see Fig. 45).[22] Immediately the Government of Canada made the vague assertion that this was "not in alignment with scientific evidence"[23] while failing to provide any evidence in support of that statement.

The fact that the polar bear has been manufactured into a symbol for fake climate catastrophe must not be challenged, not even by the people who live with the bears and must attend the funerals of their relatives and friends who are killed by them. I highly recommend this

---

19. Valerie Richardson, "University dumps professor who found polar bears thriving despite climate change," the *Washington Times*, October 20, 2019. https://www.washingtontimes.com/news/2019/oct/20/susan-crockford-fired-after-finding-polar-bears-th/.
20. Donna Laframboise, "Was this zoologist punished for telling school kids politically incorrect facts about polar bears?" *Financial Post*, October 16, 2019. https://financialpost.com/opinion/was-this-zoologist-punished-for-telling-school-kids-politically-incorrect-facts-about-polar-bears.
21. "Inuit," Wikipedia, September 10, 2020. https://en.wikipedia.org/wiki/Inuit.
22. Bob Weber, "Nunavut Draft Plan Says There Are Actually Too Many Polar Bears in Territory," *Huffington Post*, November 12, 2018. https://www.huffingtonpost.ca/2018/11/12/nunavut-draft-plan-says-there-are-actually-too-many-polar-bears-in-territory_a_23587264/.
23. Ibid.

Figure 45. Here we see the opinion of people who live among the polar bears year-round, and who have had thousands of years of experience with them. In this case "traditional knowledge" is based on personal observation, which is the basis of science.

reference from *Maclean's* magazine, a breath of fresh air in an otherwise dismal era for mass media.[24] The completed Nunavut Polar Bear Co-Management Plan was published in September 2019.[25] A thorough search on the internet found only one media report on the final adoption of the plan. It was in a newspaper published in Cambridge Bay, population 1,766, on Victoria Island, in Nunavut.[26]

The use of the polar bear as an emotionally charged symbol, often used to promote the alleged climate crisis, is one of the most classic cases of self-interest among activists seeking donations, the media seeking readers, politicians seeking votes, and scientists seeking never-ending annual grants to perpetuate the fear of extinction. As with the Great Barrier Reef and coral reefs in general, these elites are using remote iconic wildlife and natural beauty to extract financial and political support for their fraud. The general public cannot observe these

24. Aaron Hutchins, "To kill a polar bear," *Maclean's*, April 15, 2019. https://www.macleans.ca/to-kill-polar-bear/.
25. Government of Nunavut, "Nunavut Polar Bear Co-Management Plan," September 2019. https://www.gov.nu.ca/sites/default/files/nwmb_approved_polar_bear_comanagement_plan_sept_2019_eng.pdf.
26. Jane George, "Nunavut has a new polar bear management plan: NWMB," Nunatsiaq News, September 26, 2019. https://nunatsiaq.com/stories/article/nunavut-close-to-new-polar-bear-management-plan-nwmb.

claims for themselves and therefore cannot verify their accuracy. This brief passage from a *Maclean's* article speaks volumes:

> Now that communities are farther apart, and people are restricted to hunting within their own regions, polar bears roam undisturbed by humans across hundreds of kilometers of shoreline, Malliki says. 'They're increasing in numbers,' he says, but the biologists, he adds ruefully, aren't around to see them: 'Scientists live down south. They come here for a week or two and are back down again. They don't know anything about the North.'

Just when it seemed the "polar bear extinction" fraud had been finally exposed, on July 20, 2020 a multi-author paper was published in the journal *Nature Climate Change* claiming that polar bears would be at or near extinction by 2100. The lead author, Péter K. Mulnár is from the University of Toronto where he heads the Laboratory of Quantitative Global Change Ecology. His primary interest is the impact of climate change on large mammals.

The authors of this article state the aim of their study:

> Here, we establish the likely nature, timing, and order of future demographic impacts by estimating the threshold numbers of days that polar bears can fast before cub recruitment and/or adult survival are impacted and decline rapidly. Intersecting these fasting impact thresholds with projected numbers of ice-free days, estimated from a large ensemble of an Earth system model.[27]

The focus is on how long polar bears can fast before they are so negatively affected that their reproductive ability declines to the point of extinction. Clearly, a computer model is involved in predicting the "numbers of ice-free days" that might result in such a population decline. Here is another classic example of not distinguishing between ice-free days in summer versus ice-free days in winter. Polar bears come ashore in summer, but they do not "fast" unless they are hibernating. Only female polar bears hibernate, for up to five months, and that is when they give birth. The males eat what they can find in the summer including reindeer, small rodents, seabirds, waterfowl, fish, eggs, vegetation (including kelp), and berries.[28] This brings the premise

---

27. Péter K. Molnár, et al., "Fasting season length sets temporal limits for global polar bear persistence," *Nature Climate Change* 10, pp732-738, July 20, 2020. https://www.nature.com/articles/s41558-020-0818-9.
28. Emily Googheart Kautz, "What do Polar Bears Eat?" Good Nature Travel, June 15, 2019. https://www.nathab.com/blog/what-do-polar-bears-eat/.

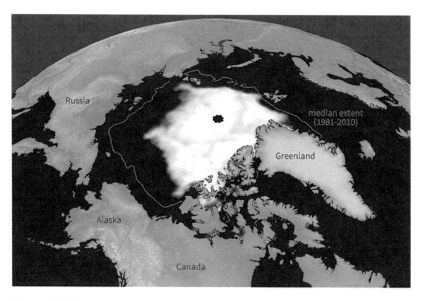

Figure 46. This figure presents sea-ice coverage from September 18, 2019, typically the time it is lowest before it begins growing again as Arctic days get shorter and colder. The red line indicates the average summer minimum from 1980 to date. The ice has receded to a similar minimum every year since 2007. This is likely beneficial for the productivity of the Arctic ecosystem.

of this paper, that polar bears fast when there is no sea ice where they can hunt seals in summer, into serious question.

The Arctic has never been known to be "ice-free" in summer (see Fig. 46). But there are many regions of the Arctic where the ice recedes completely, and polar bears must retreat to the land. The polar bears have already been proven to survive these reduced ice conditions as pointed out in Susan Crockford's research. But this paper infers that the Arctic will become completely ice-free in the winter and spring at some time between now and 2100. Only if that occurs would it be impossible for polar bears to hunt seals. The authors of this paper are once more using a computer model as if it can actually predict the future. It's time to call the alarmists out on this.

To summarize, summer ice is not really necessary for polar bear survival, but winter and spring ice is necessary for hunting seals. Looking back at history, it should be noted that if polar bears evolved as early as 350,000 years ago, they have survived through three previous interglacial periods that were warmer than this one has ever been (see Fig. 36). Even if they evolved later than deduced, they certainly lived through the Eemian interglacial period which was warmer than this Holocene interglacial we are in today. In addition, the Greenland ice core, and the ocean sediment data indicate the Earth has been in a net cooling mode for about 6,000 years and that this short period of warming we are in today is possibly just a blip on that downward trend. We know that polar bears are healthy and growing in numbers today, and that this may be due to a reduction in summer sea ice, not in spite of it. The future's not ours to see, but there is every reason to be optimistic about the climate and the survival of polar bears in the 21st Century.

CHAPTER FIVE
One Million Species Face Extinction Due to Climate Change – Soon

On May 22, 2019 Marc Morano, from Climate Depot, and I appeared along with three representatives of the awkwardly named Intergovernmental Science-Policy Platform on Biodiversity and Ecosystem Services (IPBES) before the US House of Representatives Subcommittee on Water, Oceans, and Wildlife (WOW).

The IPBES members, on behalf of the United Nations, were there to testify that their scientists had determined one million species of life are endangered, pending extinction, due to the threat of the "climate catastrophe" and the clearing of land for agricultural purposes.[1,2] Marc and I were there by invitation from the minority House Republicans to dispute this claim. We were well prepared.

There are approximately 1.74 million species on Earth that have been identified, named, and characterized. This includes one million insects; about half of which are beetles, 45,000 spiders, 305,250 other invertebrates – including shellfish, corals, nudibranchs, worms, etc.

---

1. Jeff Tollefson, "Humans are driving one million species to extinction," *Nature* 569, p171, May 6, 2019. doi: 10.1038/d41586-019-01448-4.
2. Ronald Bailey, "Leaked UN Report Says a Million Species Are at Risk of Extinction," Reason, April 26, 2019. https://reason.com/2019/04/26/leaked-u-n-report-says-a-million-species-are-at-risk-of-extinction/.

– 310,801 species of plants, 48,496 species of lichens and mushrooms, and 66,178 species of vertebrates including 34,000 fish, 7,302 amphibians, 10,038, reptiles, 10,425 birds, and about 5,513 mammals like us.[3] Note that there are many more species of insects than all other known species combined. New species are discovered regularly, in some years into the hundreds.[4] If 250 new species were discovered every year going forward it would take 27,840 years to add another 6.96 million species, bringing the count to 8.7 million.

Yet surprisingly, in 2011 we were told to believe there are actually 8.7 million species on this Earth.[5] To quote from the BBC report:

> *The natural world contains about 8.7 million species, according to a new estimate described by scientists as the most accurate ever. But the vast majority have not been identified – and cataloguing them all could take more than 1,000 years.*

The IPBES published an undated explanation on their website of how they came up with the number.[6] They simply selected a paper, written by their lead author, expanding on the conclusion that even though only 1.74 million species have been identified to date, their study "suggests" there are actually 8.7 million, plus or minus 1.3 million species.[7] As with the word "linked," the word "suggests" should never be trusted in a scientific context. These words are synonymous with "not proven," "unverified," and "guessed."

The author of this explanation is Dr. Andy Purvis, Coordinating Lead Author of the IPBES Global Assessment Report on Biodiversity and Ecosystem Services.

He is also a Biodiversity Researcher at the Natural History Museum, and a professor at Imperial College, London. Very impressive, except he mistakenly wrote in his testimony before the committee that there are 8.1 million species rather than 8.7 million which is the number in the

---

3. "*IUCN Red List of Threatened Species 2014.3. Summary Statistics for Globally Threatened Species,*" The World Conservation Union. 2014. Table 1: Numbers of threatened species by major groups of organisms (1996-2014).
4. Ashley Strickland, "*Scientists discovered 71 new species this year. Here are some of their favorites,*" CNN World, December 7, 2019. https://www.cnn.com/2019/12/07/world/new-plant-animal-species-2019-scn/index.html.
5. Richard Black, "Species count put at 8.7 million," BBC News, August 23, 2011. https://www.bbc.com/news/science-environment 14616161.
6. Andy Purvis, "How did IPBES estimate 1 million species threatened with extinction?" IPBES, undated. https://ipbes.net/news/how-did-ipbes-estimate-1-million-species-risk-extinction-globalassessment-report.
7. Camilo Mora, et al., "How Many Species Are There on Earth and in the Ocean?" *PLOS Biology*, August 23, 2011. https://journals.plos.org/plosbiology/article?id=10.1371/journal.pbio.1001127.

paper he quoted. At least it is within the "margin of error," which is plus or minus 1.3 million species for a total of 2.6 million, nearly one million more than the total number of verifiably known species. Dr. Purvis also claims in this same website posting that there are 10,000 new species discovered each year.[8] This is quite a few more than the 226 new species that were reported in 2018. Perhaps the agency should be renamed the UN Commission on Hyperbolic and Hysterical Pronouncements about Species Numbers (UNHHPSN).

The IPBES wants us to accept the number 8.7 million as a fact, despite the *true* fact that we do not have any clue as to whether the additional 6.69 million species actually exist. They say it was an "estimate described by scientists as the most accurate ever." I call this fakery, at the most shameless possible level. These species are imperceptible and have no identifications, and yet they are able to make "the most accurate estimate ever"? If something has not yet been discovered, it is impossible to "estimate" how likely its existence is, or how many of them will be discovered. Period. And the United Nations' IPBES representatives appearing before the congressional committee presented themselves as "scientists."

According to the IPBES, if one or two million of these 6.69 million unknown and unnamed species were to go extinct tonight, we would never know it happened because we didn't even know they existed in the first place. The word "logical" does not come to mind here. This is a shell game of the most cynical and dishonest variety. How did this bunch of activists, sanctioned by the United Nations, get their positions as scientists in the first place? Many questions require answers, the main one being why are we being lied to?

The Democrats on the WOW subcommittee were enthralled with the presentation. It was almost as if they were pleased to hear that one million species would go extinct unless they were able to gain administrative power and prevent the extinction by cutting off fossil fuel consumption.

Marc and I were not really expected by the Democrat majority. They clearly hoped for an uncontested hearing of the dire threat to global biodiversity caused by the climate crisis and the continued ruinous production of food for human consumption. We gave them a run for their money.

First, we pointed out that global biodiversity of species is higher now in this modern era than it has ever been in the history of life on Earth.

---

8. Andy Purvis, "How did IPBES estimate 1 million species threatened with extinction?" IPBES, undated. https://ipbes.net/news/how-did-ipbes-estimate-1-million-species-risk-extinction-globalassessment-report.

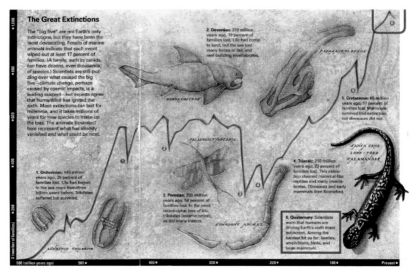

Figure 47. This graph shows the constant upward march of species diversity despite five major extinction events during the past 600 million years. The biggest event was the Permian extinction 250 million years ago when 54 percent of all taxonomic families became extinct. Yet biodiversity recovered again and again over millions of years to reach a high point in the present era.

A graph that was presented in the February 1999 issue of *National Geographic* makes this clear (see Fig. 47). Ever since multicellular life evolved more that 500 million years ago, and despite five major extinction events, the number of taxonomic families of life have constantly increased to the present day. A taxonomic family usually contains a large number of species such as the cat family, the dog family, and the weasel family in animals, and the heather family, lily family, and pine family in plants. It is one of the fascinating features of life's evolution that despite millions of species' extinctions through the ages, over time the diversity of species has increased faster than it has been reduced.

The article in which the graph appeared in *National Geographic* warned of a sixth great extinction that was apparently already underway. Unlike the previous five extinction events which were caused by asteroids or massive volcanic activity, this one is allegedly being caused by human activity, mainly climate change and the clearing of land for agriculture and urbanization. It is quite noticeable on the graph that where the line denoting taxonomic families peaks on the far left, the downturn in the line turns fuzzy. The length of that downturn indicates that about 38 entire taxonomic families have become extinct in the present era. In the bottom right there is a note referring to this drop that states:

Scientists warn that humans are driving Earth's sixth mass extinction. Among the hardest hit so far: beetles, amphibians, birds, and large mammals.

While it is a fact that humans have caused species extinction over the millennia, there is no record of any entire taxonomic family of species going extinct because of humans. When this graph was first published, I wrote to *National Geographic* and asked, "Why does the line turn fuzzy? Is it because there are actually no known taxonomic families that have become extinct in recent times? I do not know of any whole families of "beetles, amphibians, birds, and large mammals that have become extinct as implied in the text."

The reply to my inquiry came from Robin Adler, one of the researchers who worked on the article. She thanked me for "sharing my thoughts on this complicated and controversial issue" but offered no answer to my questions about the graph. Instead she asked me to:

> *Rest assured that...the many members of our editorial team...closely with numerous experts in conservation biology, paleobiology, and related fields. The concept of a 'sixth extinction' is widely discussed and, for the most part, strongly supported by our consultants and other experts in these areas, although specific details such as the time frame in which it will occur and the number of species that will be affected continues to be debated.*

I did not rest assured. No response was provided for why the line turned fuzzy. No list of extinct families of beetles, amphibians, birds, or large mammals was given. And it is abundantly clear the "experts" were not saying the sixth mass extinction was already occurring but that it was rather a prediction, by her wording, "specific details of the time frame in which it *will* occur." The article in which the sixth mass extinction was featured made it very clear that it was already well underway.[9]

It's not that human activity hasn't caused extinctions or won't cause any more in the future. The first major wave of extinctions occurred when humans came out of Africa and occupied lands where no humans had ever been. In fact, 60,000 years ago, when humans arrived in Australia, they caused the extinction of some large, slow-moving mammals that never had to escape from a human with a spear, but they certainly defended themselves against the predators that had existed there for many millions of years before.

---

9. Virginia Morell, "The Sixth Extinction," *National Geographic* (February 1999): pp42-59. It appears *National Geographic* has not archived this article. A copy of the text without illustrations has been saved here: http://163.28.10.78/content/senior/bio/tc_md/s21/ngm_Biodiversity.htm.

When humans crossed the Bering Land Bridge about 16,000 years ago during the last glaciation and occupied North America for the first time a number of large, slow-moving animals became extinct presumably due to humans hunting them for food and hides. There was a considerable pulse of extinctions beginning 500 years ago when Europeans colonized Africa, Asia, and many isolated islands such as New Zealand, Mauritius, and Hawaii. In most cases this was due to the introduction of exotic species of predators such as rats, cats, foxes, and snakes they brought with them. In other cases, it was from over-hunting, as with the extinction of the flightless, slow-moving dodo bird on Mauritius – known for being the food provisions for sailors traveling to the far east.

It is interesting to note that in equatorial Africa, where humans evolved from apes, there was little if any extinction of birds and mammals by humans. This is due to the fact that humans didn't suddenly arrive there with arrows, clubs, and spears. The animals there had evolved in tandem with human evolution for millions of years and had adapted to humans as they developed their hunting skills. Running faster than humans came in handy for potential prey.

It is completely unreasonable to compare these individual extinctions over thousands of years with the mass extinctions of the deep past. During the Permian extinction 253 million years ago, up to 96 percent of all marine species and up to 70 percent of all vertebrate land-animals became extinct.[10] It is widely believed that the five major extinctions were caused either by asteroids large enough to puncture the Earth's crust and throw billions of tons of debris into the stratosphere wherein it cut off the sunlight for the process of photosynthesis, thus in turn undermining the entire global food chain, or similarly by massive volcanic events that sent similar debris high into the atmosphere. It is considered quite certain that the most recent mass extinction, 66 million years ago, was caused by an asteroid striking northern Yucatan.[11] Named the K-T extinction, it was the cause of 75 percent of species going extinct, including all the dinosaurs, both terrestrial and marine.

At the time most of these more recent extinctions occurred, the majority of them were unintentionally brought about rather than being caused by over-hunting or eradication programs. There was simply no popular public movement in defense of those species. It was not until the extinction of the passenger pigeon in North America in 1914

---

10. Wikipedia, "Permian–Triassic extinction event," July 1, 2020. https://en.wikipedia.org/wiki/Permian%E2%80%93Triassic_extinction_event.
11. *Encyclopedia Britannica*, "K-T extinction." https://www.britannica.com/science/K-T-extinction.

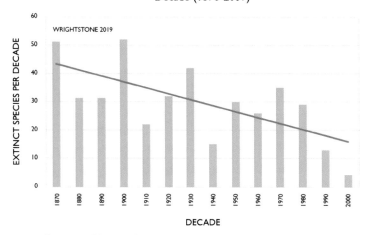

Figure 48. Graph of all species extinctions from 1870 to 2009. There is a clear downwards trend as programs were adopted to prevent species from becoming extinct. There were 413 extinctions during these 140 years. This is 0.025 percent of the 1.6 million known species today and only 0.005 percent of the alleged 8.7 million "estimated" species. It is absurd to compare this number to the five great extinctions that occurred during the past 500 million years when upwards of 50 to more than 95 percent of species went extinct in a very short period of time.

that the general public became concerned and supported programs to prevent further extinctions.[12] During the last 100 years the number of extinctions has declined by about 80 percent, largely due to efforts by naturalists, hunters, environmentalists, and politicians that gave their time to this cause (see Fig. 48).

This "species extinction" scare story is at least as preposterous as the "climate catastrophe" scare story. Both are based on unobservable factors that cannot be independently checked by sensible people. Both are supported by vested interests whose very purpose is to scare you and your family. Buyer beware.

In the end the majority Democrats on the congressional committee decided to adjourn our meeting and to convene a new meeting with a different committee some weeks later to which Marc Murano and I

---

12. Barry Yoeman, "Why the Passenger Pigeon Went Extinct," Audubon, May through June, 2014. https://www.audubon.org/magazine/may-june-2014/why-passenger-pigeon-went-extinct.

were not invited. That later meeting provided the official record for the Congress, in which the UNIPBES were uncontested and free to peddle their twaddle. Marc's and my testimony were canceled. Such is the state of scientific enquiry among congressional Democrats today.

CHAPTER SIX

The Great Pacific Garbage Patch is Full of Plastic and is Twice the Size of Texas

The Great Pacific Garbage Patch is Fabricated (and Invisible)

There seems to be no end to the absurdity spread by the "green" movement and the media. For many years the story of the Great Pacific Garbage Patch has been publicized as a great environmental tragedy. It is described as three times the size of France and twice the size of Texas. The web is full of images that give the impression that the central Pacific Ocean is completely covered in floating garbage, most of which is plastic (see Fig. 49, page 104). Here is a quote from the CNN story that covered this narrative, the footnote is below:

> A huge, swirling pile of trash in the Pacific Ocean is growing faster than expected and is now three times the size of France. According to a three-year study published in Scientific Reports Friday, the mass known as the Great Pacific Garbage Patch is about 1.6 million square kilometers in size – up to 16 times bigger than previous estimates. That makes it more than double the size of Texas.[1]

---

1. Marian Liu, "Great Pacific Garbage Patch now three times the size of France," CNN, March 23, 2018 https://www.cnn.com/2018/03/23/world/plastic-great-pacific-garbage-patch-intl/index.html.

A Part of the Great Pacific Garbage Patch

Figure 49. Images like this one are often used to depict the Great Pacific Garbage Patch.* But this image is from the aftermath of the Japanese earthquake and tsunami in 2011 that killed nearly 20,000 people, caused the Fukushima nuclear accident, and swept entire towns and villages into the sea. This photo was clearly not taken in the middle of the Pacific Ocean as there is land a short distance away in the background.

* Parsons School of Design, "The Great Pacific Garbage Patch," undated. http://b.parsons.edu/~pany468/parsons/political_website/source2/index.html.

There are also multiple photoshopped depictions of the Great Pacific Garbage Patch that claim to depict its size and location (see Figs. 50 and 51). Many people believe these represent the truth as they are not able to go to the middle of the Pacific Ocean to check for themselves. But these images are 100 percent fabricated.

Fortunately, there is a very clear image of the entire Pacific Ocean which is also a composite taken over a year-long period in order to create a clear and perfectly cloudless image. It does not show any visible garbage patch because there isn't one. The Great Pacific Garbage Patch, famed to be twice the size of Texas is fabricated, as it does not actually exist (see Fig. 52, page 106).[2]

---

2. Sarah Knapton, "'Great Pacific Garbage Patch' is a myth, warn experts, as survey shows there is no 'rubbish island,'" the *Telegraph*, October 5, 2016. https://www.telegraph.co.uk/science/2016/10/05/great-pacific-garbage-patch-is-a-myth-warn-experts-as-survey-sho/.

Figure 50. This illustration of the Great Pacific Garbage Patch indicates that it is nearly as large as the entire lower 48 US states. The website it appears on claims this is the result of "new aerial surveys" yet it also states, "the patch is not truly visible with the naked eye."*
* Mihai Andrei, "Aerial survey shows the 'Great Pacific Garbage Patch' is much larger than we thought," ZME Science, October 6, 2016. https://commons.wikimedia.org/wiki/File:Pacific_Ocean_satellite_image_location_map.jpg.

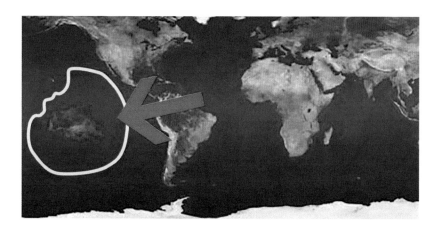

Figure 51. This depiction of the Great Pacific Garbage Patch shows it to be about half the size of South America. This is actually a composite of many satellite images which were stitched together so it could exhibit nearly the entire planet without cloud cover. The Garbage Patch was simply painted or photoshopped onto the photograph. The website this appears on says the garbage patch is a "satellite image."*
* Isabella Pruna, "Ocean Health and Advocacy," November 18, 2018. http://art3170csula.blogspot.com/2018/11/ocean-health-and-advocay-isabella-pruna.html.

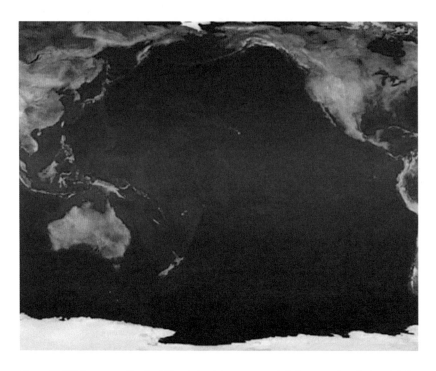

Figure 52. This is a composite of satellite photos taken over a period of one year in order to catch the entire area with no cloud coverage. The Hawaiian Islands can be easily seen, and they are definitely not twice the size of Texas.

I have been confronted with some hostility among members of my audiences when I point out the fact that the garbage patch is a hoax. The most common rebuttal is that the garbage patch is just below the surface and that is why it can't be seen from a satellite or an aircraft. In my experience, and I have spent a lot of time scuba diving and snorkeling, most objects either float or sink and only rarely do they hover at a particular depth below the surface. Plastic articles do not have buoyancy-compensation devices, they are mostly either less dense or more dense than the water they are in. One person even stated that the reason you can't see the Great Pacific Garbage Patch is because it's only the clear plastic.

As a final resort, the garbage-patch defenders claimed that the plastic is in the form of micro-plastics in the water column, or in other words, they're invisible. Just another garden-variety invisible, fake catastrophe like so many others. At this point one can only shake their head and call it off. If the plastic consists of invisible micro-plastic, how can it be labeled the Great Pacific Garbage Patch twice the size of Texas?

Figure 53. This ship spent 48 days combing the "Pacific Garbage Patch" between California and Hawaii and salvaged 103 tons of plastic. One can see that nearly all of it is discarded fishing gear, mostly nets and ropes. Fishnets are a serious problem, not because they are plastic but because they are meant to catch fish and other marine life. They are referred to as "ghost nets" because they can entangle marine life. A program focused on educating fishers would go a long way to reducing this problem.*

* Jessica Stewart, "Largest Ocean Cleanup Hauls 103 Tons of Plastic from the Pacific Ocean," My Modern Met, July 8, 2020. https://mymodernmet.com/ocean-voyages-institute-great-pacific-garbage-patch-cleanup/.

Of course, there is plastic debris floating in the oceans, but it is not in a giant patch. It is very spread out and most of it is discarded fishing gear (see Fig. 53).

Is Marine Plastic an Environmental Catastrophe?

A tremendous effort has been made and is still underway to make "plastic" into a negative term. This is similar to the campaigns against fossil fuels which literally refer to them as "dirty coal" and "dirty oil." Clearly there is nothing evil about dirt as dirt is where we grow most of the food we eat and the food we feed to our livestock. In the case of fossil fuels the word "dirty" is simply being used as a pejorative, or derogatory slur, as in "dirty rotten scoundrel." It is propaganda, one of the main features of which is associating the object of your disdain with negative words like "dirty."

The "War on Plastic" is, in fact, a proxy for the campaigns against fossil fuels. Nearly all the synthetic plastic polymers are made from fossil fuels. Polystyrene and polypropylene are made from oil. Polyethylene can be made from oil or natural gas. Polyvinylchloride (PVC or vinyl) is made from natural gas and table salt (NaCl).

Interestingly, nature produced a plastic polymer through evolution more than three billion years ago. Unbeknown to most people, cellulose is a polymer of glucose, which in turn is the main sugar produced by combining CO_2 and H_2O in photosynthesis. Early life forms evolved the ability to join glucose molecules in a string to form the fibers for building cell walls. So cellulose was the first plastic, and, in that sense, paper is a plastic as it too is composed of cellulose. Rayon is a plastic because it is synthesized from cellulose. Wood is also technically a plastic as it is composed of cellulose and lignin, both of which are used to make synthetic polymers and plastics. Cotton is another plastic as it is almost pure cellulose. Plastic is not a dirty word. Plastic is one of the most versatile and useful groups of materials. Life without plastics like cellulose would simply not be possible.

The anti-plastic movement has been very successful at portraying plastic as extremely negative from an environmental perspective. One of the most effective tools in this portrayal are the staged and contrived images that are intended to make people cringe at the sight of them (see Fig. 54).

In the same article with the photo of a dead bird full of plastic, there is an account of 13 sperm whales stranded on a beach in Germany. The article claims that their stomachs were full of plastic. However, the article provides no evidence of this either with images or citations.[3] The only photo of whales in the article showed two of the 13 sperm whales stranded on a beach (see Fig. 55).

Here is another entirely staged image used to demonstrate that whales are swallowing huge amounts of plastic too. The whale is artificial, and the plastic has been shoveled into its mouth by Greenpeace members in the Philippines (see Fig. 56, page 110).

The Many Benefits of Marine Plastic

It is true that plastic can cause harm in the oceans as with the example of discarded fishnets continuing to catch fish and entangling other marine species such as turtles and even whales. Convincing fishers to bring their damaged nets back to the dock for disposal should be the

---

3. Alexander Haro, "13 Sperm Whales Found Dead with Stomachs Full of Plastic Trash," the Inertia, June 3, 2016. https://www.theinertia.com/environment/13-sperm-whales-found-dead-with-stomachs-full-of-plastic-trash/.

Figure 54. This dead albatross chick has been staged by cutting it open and stuffing it with plastic objects. When birds swallow plastic it is directed to the gizzard. Unless this bird had a gizzard the size of its entire body this is a fake photo. But it does have the desired effect: revulsion.

Figure 55. Two of the 13 sperm whales that were stranded on a beach in Germany in January 2016. Whale strandings are quite common with hundreds of whales around the world stranding annually. This has been happening for millions of years, long before modern plastics were invented.*
* Ed Yong, "The Tiny Culprit Behind A Graveyard of Ancient Whales," February 25, 2014. https://www.nationalgeographic.com/science/phenomena/2014/02/25/the-tiny-culprit-behind-a-graveyard-of-ancient-whales/.

Figure 56. A mock whale with its mouth full of plastic is the only image associated with an article about a sperm whale that washed up in Spain that allegedly had 64 pounds of plastic in its stomach. No evidence of the actual sperm whale was presented in the article.*

* Chris McDermott, "Giant 'Dead Whale' Is Haunting Reminder of Massive Plastic Pollution Problem," EcoWatch, May 15, 2017. https://www.ecowatch.com/dead-whale-plastic-pollution-2408402292.html.

aim of a major international campaign in itself. And it is possible that whales sometimes swallow plastic. But there is no indication that this is causing the kind of mass death that is implied in these articles.

It was mentioned earlier that wood is a kind of plastic, made with hydrogen and carbon like synthetic plastic, except wood also contains oxygen. There is a lot of wood floating in the ocean. Here is a description from an environmentalist about the ecological role of wood in the sea.

> For driftwood that leaves terra firma to begin a new life at sea, the odds of ever returning to land are pretty slim. But being lost at sea doesn't necessarily mean their travels are a lost cause. As writer Brian Payton noted recently in Hakai *magazine, driftwood can stay afloat in the open ocean for about 17 months, where it offers rare amenities like food, shade, protection from waves, and a place to lay eggs. As such, pelagic driftwood becomes a 'floating reef' that can host a variety of marine wildlife.*
>
> *That includes wingless water striders (aka sea skaters), which lay their eggs on floating driftwood and are the only insects known to inhabit the open ocean. It also includes more than 100 other species of invertebrates, Payton adds, and some 130 species of fish.*[4]

4. Russell McLendon, "The Surprising Beauty and Benefits of Driftwood," Treehugger, August 15, 2018. https://www.treehugger.com/driftwood-beauty-benefits-4869726.

Figure 57. The following caption was written under this photo of a crab in a plastic cup. "A crab was trapped inside a discarded milktea cup in Verde Island, Philippines (© Noel Guevara / Greenpeace)."* But in fact, the crab is not "trapped." It is using the cup as shelter, in other words, its habitat. Hundreds of marine species benefit from plastic objects whether they are floating, drifting in the water column, or lying on the seafloor.

* Greenpeace International, "Learn About Plastic Pollution," 2020. https://www.greenpeace.org/international/campaign/toolkit-plastic-free-future/learn-about-plastic-pollution/.

From the point of view of all these species that use driftwood as habitat, floating plastic is no different. As with driftwood, drifting plastic is a floating reef that provides a home for many marine species. In fact, plastic offers much more variety in terms of shapes such as bottles and containers, so it offers a wider range of habitats than driftwood (see Fig. 57).

Plastic is no more toxic than driftwood, which is not toxic at all. Yet website after website claims that plastic "leaches toxics" and "chemicals" into the oceans.[5,6] This is untrue. There is a good reason why we package and wrap much of our food in plastic containers and plastic wrap. It is because the plastic protects it from contamination and spoilage, and because the plastic is sterile and does not contain anything toxic. But wait a minute, the plastic polyvinylchloride contains chlorine which is an elemental gas and is very toxic. What about that? The fact is that the most common salt in seawater is sodium chloride,

---

5. Hannah De Frond, et al., "Plastic Pollution is Chemical Pollution," Ocean Conservancy, April 23, 2019. https://oceanconservancy.org/blog/2019/04/23/plastic-pollution-chemical-pollution/.
6. Hadley Leggatt, "Toxic Soup: Plastics Could Be Leaching Chemicals into Ocean," *Wired*, August 9, 2019.

Figure 58. Pelagic gooseneck barnacles have engulfed a small plastic fishing float before it was turned from flotsam into jetsam. If a piece of plastic is floating in the sea it will not be long before it becomes habitat for a potential host of species (Photo – Patrick Moore).

otherwise known as table salt. Sodium chloride is an essential nutrient for all animal life including marine life.[7] Isn't chemistry fascinating?

One of the most important benefits of plastic in the sea is that, like driftwood, it provides a habitat for many of the species such as barnacles that attach themselves to floating objects (see Fig. 58). Other species lay their eggs on driftwood and floating plastic. This in turn provides food for fish and birds. There is no doubt that the benefit of additional food provided by floating plastic far outweighs the rather rare occurrence of damage or death from being tangled in plastic.

While on this subject it is important to note that the vast majority of animals are in the shape of a tube. This is true for all vertebrates, worms, clams, insects, shrimp, sea cucumbers, and so many more, including humans. Food is ingested through one end of the tube and waste is emitted from the other end. Evolution has been intelligent enough to make the in-hole smaller than the out-hole. Therefore, nearly anything you can swallow – without choking to death – will be able to pass through the rest of you. Regarding humans, the Harvard

---

7. Exploring Our Fluid Earth, "Weird Science: Salt is Essential to Life," University of Hawaii, 2020. https://manoa.hawaii.edu/exploringourfluidearth/chemical/chemistry-and-seawater/salty-sea/weird-science-salt-essential-life.

Medical School's website on bowel obstructions lists numerous causes but there is no mention of large objects that were swallowed.[8]

During my numerous deep-sea voyages on Greenpeace campaigns, I witnessed many occurrences of plastic with sea-life growing on them. There is no doubt that one of the reasons fish and birds sometimes ingest bits of plastic is to get to the food growing on it. They would do the same thing if the food was growing on a small piece of wood.

One of the most contrived "narratives" about plastic in the ocean is that adult albatrosses are feeding plastic bags to their chicks and that in some cases this is killing the chicks. In his series *Blue Planet* for the BBC, Sir David Attenborough displays plastic bags and other bits of plastic film which he claims was fed to albatross chicks by their parents.[9] There is no video or photos of the parent albatross actually feeding plastic bags to their chicks.

Unlike mammals, birds have no teeth and therefore cannot chew their food. Birds of prey such as eagles, ospreys, and hawks are able to rip large prey into pieces small enough to swallow. But most birds swallow their food whole. Unlike most mammals, birds have two stomachs: one like ours with gastric acids to begin digestion and then there's the muscular gizzard where the food is broken down with the aid of indigestible solid objects. Birds that live on or close to land use pebbles for this. However, there are not a lot of pebbles out at sea. For sea birds the first choice is pumice from undersea volcanos which are basically rocks that float. When pumice is scarce seabirds will use pieces of hard wood, floating nuts from trees, and anything else that is the right size and relatively hard. In addition, the hard beaks of squids eaten by seabirds are retained in the gizzard to help with digestion. Since plastic was introduced to the ocean, seabirds are perfectly content to use it as one of the hard objects for their gizzard. They are not giving bits of plastic to their chicks because they mistake it for food, birds are not that stupid. They are giving it to them so they can digest their food.10

Sir David Attenborough is the author of The Life of Birds, which he published in 1998 based on the 10-part PBS series.11 Is it possible that he does not know about gizzards in birds? Not to mention the fact that

---

8. Harvard Medical School, "Bowel Obstruction – What is it?" Harvard Health Publishing, February 2020. https://www.health.harvard.edu/diseases-and-conditions/bowel-obstruction-a-to-z.
9. "Albatrosses are ingesting plastic," *Blue Planet II*: Episode 7 Preview – BBC One. https://www.youtube.com/watch?v=I4QNoIP7Khc.
10. *"Digestion,"* Fairbanks Science Center, undated. http://www.fernbank.edu/Birding/digestion.htm#:~:text=Birds%20have%20a%20two%20part,structure%20of%20the%20food%20material.
11. David Attenborough, *The Life of Birds*, Princeton University Press, October 18, 1998. https://www.booktopia.com.au/life-of-birds-sir-david-attenborough/book/9780691016337.html.

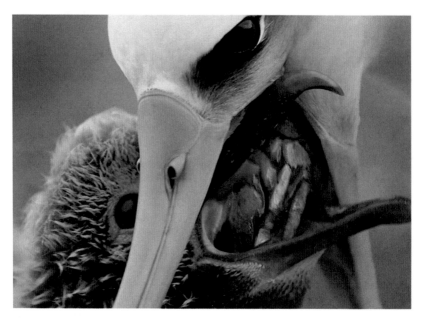

Figure 59. An albatross parent transfers suitable bits of hard plastic to its chick. These will be used to aid digestion in the gizzard. For seabirds it is more difficult to find suitable hard objects than for birds on land which use pebbles. The addition of floating plastic bits of marine flotsam has been a benefit to seabirds when other suitable objects are hard to find.

chicks need hard objects fed to them in order to digest their food properly; and that they continue to ingest hard objects through their entire adult lives for the very same purpose? Why doesn't Attenborough even mention the word "gizzard" during his claim that albatross parents are feeding plastics to their chicks?

The only image I was able to find in a thorough internet search shows an albatross transferring bits of hard plastic to its chick, suitable for its function in the gizzard, not plastic bags (see Fig. 59).[12]

On the website hosted by the Ocean Portal of the Smithsonian and the National Museum of Natural History, this claim is made:

> *Many birds accidentally eat plastic and other marine debris floating in the ocean, mistaking it for food. But the problem is intensified in Laysan albatrosses because of the way they catch fish, squid, and other seafood: by skimming the surface of the water with their beak. Along the way, they accidentally pick up a lot of floating plastic, which they then feed to their chicks.*

---

12. Albatross. https://www.albatrossthefilm.com/watch-albatross.

> Adults can regurgitate plastic they've swallowed, but chicks are unable to, so it fills up their stomachs.

> The effects of plastic on the chicks hasn't been scientifically proven. It's probable that it injures or kills the birds by cutting their stomachs.[13]

The Smithsonian should be ashamed of itself for spouting these false claims. As pointed out above, one of the reasons birds and fish swallow bits of plastic is to get to the food that is growing on them. Birds are not so stupid as to "mistake" plastic for food, or to feed plastic bags to their chicks thinking it is food. But most seabirds, including albatross, do purposefully provide hard bits of plastic to their chicks, and as adults they ingest plastic bits themselves, for the same reason that birds on the land feed pebbles to their chicks and ingest pebbles themselves for their entire lives.

The Smithsonian statement is not credible in that it states: "Adults can regurgitate plastic they've swallowed, but chicks are unable to, so it fills up their stomachs."

This is a false statement. While albatross chicks are in the nest, they are given as much as a kilo of hard objects to assist with digestion in their gizzard. Before they fledge and take flight, they regurgitate most of the hard objects in their gizzard, otherwise they would be too heavy to fly. The material they cough up is called a "bolus." Take note that the Smithsonian does not mention the words "gizzard" or "bolus." Like Sir David Attenborough, they intentionally neglect to divulge this information regarding the existence of the gizzard, even though they are most certainly aware of its existence. They and their colleagues are lying to us, knowing that the average citizen cannot go to remote ocean islands to observe the truth for themselves. This is a crime against the public. Scientists should be held to a high moral standard, and this sort of propaganda does not even qualify for a low moral standard. On the same websites there is a collection of images exhibiting dead albatross chicks cut open to reveal what the Smithsonian alleges are the contents of the bird's "stomachs." I believe these images are staged (see Fig. 60, page 116). I stand to be corrected but I do not find these images to be credible as the amount of plastic shown is at least ten times as much as has been documented in the gizzards of albatross chicks.

---

13. The Ocean Portal, "Laysan Albatrosses' Plastic Problem," Smithsonian, undated. https://ocean.si.edu/ocean-life/seabirds/laysan-albatrosses-plastic-problem#:~:text=Along%20the%20way%2C%20they%20accidentally,it%20fills%20up%20their%20stomachs.

Figure 60. This is one of the images on the Smithsonian's website which does not make mention of the fact that albatrosses do, in fact, use hard bits of plastic for aiding in the digestion of their food. It is not because they are mistaking the plastic for food, it is because all birds have a gizzard where they use suitable hard objects to help their digestion. These images are widely distributed throughout the internet and are used to give plastic a bad name, when in fact the bits of plastic are serving a very useful purpose. These images are almost certainly staged as no albatross chick has this much plastic in its gizzard. All the studies researched for this chapter indicate that plastic objects are a minority of the objects in albatross chicks' gizzards.

Greenpeace is right in the center of this disinformation campaign. Below is an excerpt from a paper describing Greenpeace's campaign policy and fundraising strategy that I co-authored with Dr. Willie Soon, as well as with Drs. Michael, Ronan, and Imelda Connolly.[14]

(Begin Excerpt)
Greenpeace claim that the Laysan albatross is in danger of extinction and they are blaming plastics. Here is an excerpt from Greenpeace USA's webpage on the albatross:

> *An even more tragic cause for albatross mortality is consumption of marine debris, mainly plastic, that they mistake for food. Birds are found with bellies full of trash, including cigarette lighters, toothbrushes, syringes, toys, clothespins, and every other type of plastic material. On Midway Atoll, 40 percent of albatross chicks die due to dehydration and starvation from trash filling their*

---

14. Michael Connolly, et al., "The Truth Behind the Plastic 'Crisis,'" Climate Depot. December 14, 2018. https://www.climatedepot.com/2018/12/14/special-report-scientists-expose-the-truth-behind-the-plastic-crisis-greenpeace-co-founder-the-sea-of-plastic-is-a-fiction-the-ultimate-in-fake-news/.

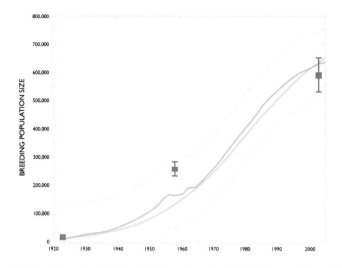

Figure 61. A graph of the population of Laysan Albatross.* Their population was decimated by feather hunters,** but they have recovered steadily and may eventually reach their carrying capacity where the population will level off. Clearly, the great "menace" of marine plastic has not slowed them down. Greenpeace claims that the species is "in real danger of extinction because they are unable to breed fast enough to keep up with population declines."*** Perhaps they have not seen this graph.
* Javier A. Arata, "Status Assessment of Laysan and Black-Footed Albatrosses, North Pacific Ocean, 1923-2005," US Geological Survey, 2009. https://pubs.usgs.gov/sir/2009/5131/pdf/sir20095131.pdf.
** "The Feather Trade," New Hampshire PBS, 2020. https://nhpbs.org/wild/feather.asp.
*** Greenpeace, "Albatross," December 2015. https://www.greenpeace.org/usa/oceans/wildlife-facts/albatross/.

bellies providing no nutrition. It has been estimated that albatross feed their chicks about five tons of plastic a year at Midway Atoll.

And:

This species is in real danger of extinction because they are unable to breed fast enough to keep up with population declines (see Fig. 61).[15]

It has been known since the 1960s that many seabird species routinely ingest pieces of plastic as well as other indigestible objects such as squid beaks, pumice stones (a type of floating volcanic rock), nuts, wood, and other floating objects.

More than 40 years later, Gray, et al. (2012), carried out a similar

---

15. Greenpeace, "Albatross," December 2015. https://www.greenpeace.org/usa/oceans/wildlife-facts/albatross/.

analysis of a sample of 18 Laysan albatrosses and 29 Black-footed albatrosses, except focusing specifically on the plastic material. Their study was published in the journal *Marine Pollution Bulletin*.[16]

When researchers first started noticing, in the 1960s through the 1980s, that seabirds were ingesting plastic particles, they were shocked and alarmed. The first thought was that the seabirds were "eating" the plastic by mistake. Maybe the seabirds thought the plastic was "food"? Or maybe they were accidentally swallowing the plastic along with food?

More alarmingly, seabird chicks seemed to have more than 10 times as much plastic in their stomachs as the adults. It seemed that the seabird mothers were feeding their chicks even more plastic particles than the adults were ingesting.

Researchers became worried that all of this plastic might be harming the seabirds. In particular, they were most worried about the following:

1. Could sharp plastic fragments cut the stomach linings of the birds, causing them to starve and die?
2. Could the build-up of this undigested plastic in their stomachs give them a false sense of being "full" and cause them to starve?
3. Could the "extra" plastic in their stomachs make it harder for the chicks and adults to fly?

These were all reasonable concerns that led to a lot of research. As we will discuss later, after all of this research was concluded, the answers to all of the above now seem to be a simple "no." Not only do the seabirds seem to be doing fine with these small plastic particles, but seabirds seem to be intentionally seeking them out as a useful digestive aid. The plastic seems to be a beneficial alternative to the naturally occurring pumice, squid beaks, and other hard indigestible objects that seabirds have been using as digestive aids for millions of years.

Partly because they are a relatively large bird, the albatrosses are the seabirds which seem to ingest the largest amount of plastic. But many other seabird species also ingest small quantities of plastic particles along with other indigestible material (squid beaks, sand, insects, etc.). Moser & Lee (1992) carried out one of the most comprehensive long-term surveys over the period 1975 to 1989. In total, they analyzed 1,033 seabirds, making up 38 species. They found that 21 of the 38

---

16. Holly Gray, et al., "Incidence, mass and variety of plastics ingested by Laysan (Phoebastria immutabilis) and Black-footed Albatrosses (P. nigripes) recovered as by-catch in the North Pacific Ocean," *Marine Pollution Bulletin* 64 (2012) pp2190-2192. http://oikonos.org/wp-content/uploads/2015/02/Incidence-mass-and-variety-of-plastics-ingested-Gray-2012.pdf.

species had ingested at least some plastic, and some species (Northern Fulmars, Red Phalaropes, and Greater Shearwaters) frequently ingested plastic particles.[17]

So, it is true that many seabirds are ingesting plastic particles along with many other types of naturally occurring indigestible materials. The average size of these plastic particles is very small (less than one gram). The plastic only comprises a small fraction of the indigestible material the seabirds ingest – naturally occurring pumice and squid beaks seem to be more popular. Also, seabirds seem to have been doing this since the 1960s, and if we compare the results of Kenyon & Kridler[18] and Gray, et al.,[19] there doesn't seem to have been an increase in the amount of plastic being ingested since then.

However, Greenpeace has chosen to ignore all of this research, and instead (falsely) insists that the answer to all of the above concerns is "yes." Further, they basely imply that this is a new and growing "crisis" and that it is somehow related to the developed world's usage of "single use plastics":

> *Our oceans are slowly turning into a plastic soup and the effects on ocean life are chilling. Big pieces of plastic are choking, and entangling turtles and seabirds and tiny pieces are clogging the stomachs of creatures who mistake it for food, from tiny zooplankton to whales. Plastic is now entering every level of the ocean food chain and even ending up in the seafood on our plates.*[20]

> *The plastics obstruct the animals' intestines, block gastric enzyme secretion, and there are growing fears that they might also disrupt hormone levels or cause other biological effects as a result of the chemical burden they carry. It is estimated that up to about one million seabirds and 100,000 marine mammals die each year from ingesting plastic or by getting tangled in nylon fishing line, nets, six-pack plastic can holders, and plastic rope.*[21]

---

17. Mary L. Moser, et al., "A Fourteen-Year Survey of Plastic Ingestion by Western North Atlantic Seabirds," *Colonial Waterbirds* Vol. 15, No. 1 (1992), pp83-94. https://www.jstor.org/stable/1521357?origin=crossref&eq=1.
18. Karl W. Kenyon, et al., "Laysan Albatrosses swallow indigestible matter," the *Auk*, Volume 86, Issue 2, April 1, 1969, pp339-343. https://academic.oup.com/auk/article-abstract/86/2/339/5209529?redirectedFrom=PDF.
19. Holly Gray, et al., "Incidence, mass and variety of plastics ingested by Laysan (Phoebastria immutabilis) and Black-footed Albatrosses (P. nigripes) recovered as by-catch in the North Pacific Ocean," *Marine Pollution Bulletin* 64 (2012) pp2190-2192. http://oikonos.org/wp-content/uploads/2015/02/Incidence-mass-and-variety-of-plastics-ingested-Gray-2012.pdf.
20. Greenpeace UK, "Plastic Pollution," undated. https://www.greenpeace.org.uk/challenges/plastic-pollution/.
21. Rex Weyler, "The Ocean Plastic Crisis," Greenpeace International, October 15, 2015.

If any of Greenpeace's scary-sounding claims were true, then it would be a cause for concern. But, let us look at what the scientists who were actually investigating the claims have concluded.

One of the first systematic efforts to investigate these claims was through the graduate work of Peter Ryan in the late 1980s and early 1990s. Although he considered each of the proposed mechanisms by which the plastic particles might potentially be harming the seabirds, all of the evidence suggested that the seabirds were doing fine. A good summary of his findings is provided in a presentation he gave to the 1989 International Conference on Marine Debris in Hawaii:

> *Few statistically significant negative correlations have been found among adequately controlled samples, suggesting that the effects of ingestion are either relatively minor or that they frequently are masked by other variables.*[22]

Other researchers' findings were the same. For example, here is the main conclusion from the Moser & Lee study which we mentioned above:

> *We found no evidence that seabird health was affected by the presence of plastic, even in species containing the largest quantities: Northern Fulmars (Fulmarus glacialis), Red Phalaropes (Phalaropus fulicaria), and Greater Shearwaters (Puffinus gravis)."*[23]

Moser and Lee also directly addressed several of the concerns that researchers had originally raised. With regards to the claim that seabirds were starving because they were mistakenly feeling "full" from the plastic, they found the claim was wrong:

> *Plastic ingestion may cause seabird starvation if the presence of plastic in the gut signals satiety and reduces bird appetites [quoting a 1985 study suggesting this might be a problem]. We found no evidence for this effect in the seabirds analzsed in this study. Stomach fullness was not correlated with the amount of plastic in the gut.*[24]

---

22. Peter Ryan, "The effects of ingested plastic and other marine debris on seabirds," in R. S. Shomura, et al., "Proceedings of the Second International Conference on Marine Debris" April 2-7, 1989, Honolulu, Hawaii, volume 1. NOAA Technical Memorandum, NMFS-SWFSC (154): pp623-634. https://pdfs.semanticscholar.org/9654/5749c2085c0a9e8c62143e9dc6ebec6cdee3.pdf.
23. Mary L. Moser, et al., "A Fourteen-Year Survey of Plastic Ingestion by Western North Atlantic Seabirds," *Colonial Waterbirds* Vol. 15, No. 1 (1992), pp83-94. https://www.jstor.org/stable/1521357?origin=crossref&seq=1.
24. Ibid.

They also considered the claim that the plastic might be cutting the stomach linings of the birds. Again, they found no evidence for this claim:

> *Although approximately 20 Northern Fulmars and Greater Shearwaters in our collection had plastic accumulations large enough to potentially alter gizzard function, we found no evidence of digestive tract impaction or occlusion.*[25]

They agreed that plastic ingestion was widespread among seabirds, but the worries that this was harming them seem to have been unfounded:

> *In conclusion, our results indicated that plastic ingestion is widespread among western North Atlantic seabirds...However, we found no evidence that plastic particle ingestion is detrimental to western North Atlantic seabirds. The species most likely to suffer health risks from ingestion of ocean-borne plastics, Northern Fulmars, Red Phalaropes, and Greater Shearwaters, showed no ill effects, with Northern Fulmars actually increasing their abundance and range in the western North Atlantic during the study period.*[26]

Another potential concern was that the plastic particles were gradually building up in the seabirds' stomachs, and that over time their stomachs would become filled with plastic. However, it is now becoming clear that, like the other indigestible material that seabirds ingest (squid beaks, pumice, etc.), the plastic particles only last for a few months before being worn down.

For instance, van Franeker and Law (2015) found that some of the early estimates for the length of time the plastic remained in the stomachs (six months to a year or even longer) were too long. They found:

> *An overall 90 percent decrease in the average number of plastic particles in the stomach over summer from 8.6 particles/bird in May, to 3.2 in June, 1.2 in July, and 0.8 in August.*[27]

---

25. Ibid.
26. Ibid.
27. Jan A. van Franeker and Kara Lavender Law, "Seabirds, gyres and global trends in plastic pollution," *Environmental Pollution*, Volume 203, August 2015, pp89-96. https://www.sciencedirect.com/science/article/pii/S0269749115001104?via%3Dihub.

It appears that rather than the plastic particles "building up" in the seabirds' stomachs, the seabirds actually have to collect replacement particles every year.

So, despite Greenpeace's repeated claims, we now know that the ingestion of plastic particles by seabirds doesn't seem to be having any ill effects on the birds.

(End of Excerpt).

It is clear that Sir David Attenborough, the BBC, the Smithsonian, and Greenpeace are knowingly lying through their teeth in hopes no one will call them out. Very few people know the truth about seabirds, gizzards, and the hard objects used for digestive aids. Virtually no one except scientists do field research in the breeding grounds of seabird species. And activists, the media, and some scientists are happy to play along with the "birds mistaking plastic for food" narrative. This is a clear example of an unobservable, remote situation where the general public cannot verify the truth of the matter for themselves.

Surely the concern about plastic in the oceans should be narrowed to focus on discarded fishnets and other articles that can trap or injure marine species. In the final analysis, the multi-faceted benefits of plastic far outweigh the negatives.

Henderson Island – "The Most Plastic-Polluted Place on Earth"

Not many people know where Henderson Island is situated because it is one of the most remote bits of land in the world. To get to it one must first fly to Tahiti, and then fly another four-five hours to Mangareva Island in the Gambier Islands, and then a three-day boat trip to Pitcairn Island, and then you must find someone with a boat who will take you the final 168 kilometers (104 miles) to Henderson Island which is uninhabited, and has no dock. It is just south of the Tropic of Capricorn.

On June 6, 2017, the journal *Proceedings of the National Academy of Sciences of the United States of America* published a paper that claimed "The density of (plastic) debris was the highest recorded anywhere in the world," on Henderson Island.[28] This was followed by a large number

---

28. Jennifer L. Lavers, et al., "Exceptional and rapid accumulation of anthropogenic debris on one of the world's most remote and pristine islands," PNAS, June 6, 2017, 114 (23) pp6052-6055. https://www.pnas.org/content/114/23/6052.

of media reports with photos claimed to be taken at Henderson Island (see Figs. 62 and 63).[29,30]

It takes weeks and tens of thousands of dollars at the best of times to visit Henderson Island. With the COVID-19 restrictions it may be impossible. So, I zoomed in on it with Google Earth Pro (see Figs. 64, 65, and 66). Here is the entire island, 37.3 square kilometers (14.4 square miles).

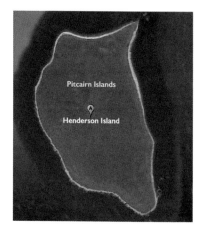

Figure 62. Henderson Island is a raised coral reef. There is no fresh water supply as the rain soaks right though the porous coral. It is uninhabited. It was designated a World Heritage Site by the United Nations in 1988.

Figure 63. One of many photographs claiming to be of beaches on Henderson Island. Wherever it actually is, the plastic is mostly fishing gear.

---

29. Deborah Byrd, "This Pacific island is the most plastic-polluted place on Earth," *EarthSky*, April 22, 2018. https://earthsky.org/earth/henderson-island-pacific-plastic-pollution-study.
30. Kaushik Patowary, "Henderson Island: This Uninhabited Island is the World's Most Polluted," *Amusing Planet*, March 24, 2018.

It is claimed, in the various reports about plastic debris, that the most polluted beaches are on the eastern shore. Here are some images of the sandy beaches on the eastern shore. I zoomed in on every beach on the island and did not see a single thing that looked like plastic debris. I believe at this resolution, if there were the amount of plastic on the beaches that their photographs display as the current extent of pollution, it should certainly be possible to see it (see Figs. 64, 65, and 66).

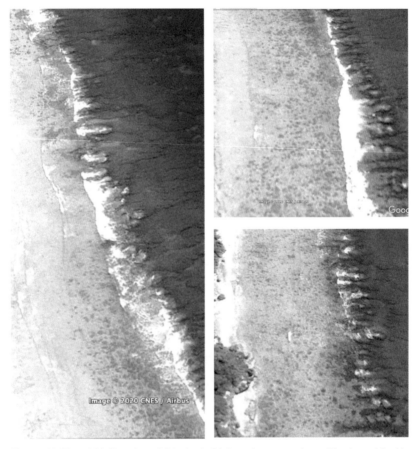

Figures 64, 65, and 66. There is no visible plastic debris on the eastern shore of Henderson Island in these images. I zoomed in on every beach on the island and they all appear similar to these stretches of beach. An independent team should go to Henderson Island to verify the claim that the Island, in fact, has the most plastic debris of anywhere else on Earth.

Waste-to-Energy
The Solution for Unwanted Combustible Materials

While on the subject of plastic waste, let's cast a wider net and consider the subject of garbage in general, usually referred to as municipal solid waste. Waste materials can be divided into three broad categories; metals, glass, and combustibles (paper, wood, plastic, and food refuse). Metals are among the easiest and most economical materials to recycle. In the United States 86 percent of steel is recycled. Glass is also easily recycled, and any surplus or inferior product can be used as aggregate in asphalt or concrete. Most paper products are recyclable and a very high percentage of them are recycled today, however, paper that is contaminated with grease or food cannot be recycled. Wood waste, especially from the demolition of buildings, is more problematic as it is often discarded in short pieces and full of nails. A good amount of plastic can be recycled, though a large percentage of the material – plastic film and contaminated plastics – is not recycled. Food waste can either be converted into waste-to-energy fuel or it can be composted for garden soil. As a result, most waste that ends up in landfills is composed of the combustible materials that could actually be used to produce electricity and heat. This truly is a waste of valuable resources.

Interestingly, the most common combustible materials are of life origin. Wood and paper are made from trees and plastics are made primarily from oil and natural gas, both of which are transformations of sediments from marine life; and of course food waste is of life origin as well. These carbon-based combustible materials are ultimately the product of solar energy and photosynthesis. We can recover that energy and turn it into electricity and heat.

The ultimate solution to preventing the dumping of unwanted combustible waste into a landfill or littering the environment is an industrial waste-to-energy plant (also called an energy-from-waste plant). This has been proven in cities and counties around the world. The technology is state-of-the-art with pollution control that meets the rigid standards adopted in most industrialized countries (see Fig. 67, page 126).

There are more than 2,450 waste-to-energy plants worldwide and there are plans to build more than 1,100 more plants in the near future.[31] The global market for these plants is expected to grow by 6.5 percent annually from 2020 to 2025. The European Union has 492

---

31. Ecoprog, "Waste to Energy 2019/2020," undated. https://www.ecoprog.com/publications/energy-management/waste-to-energy.htm.

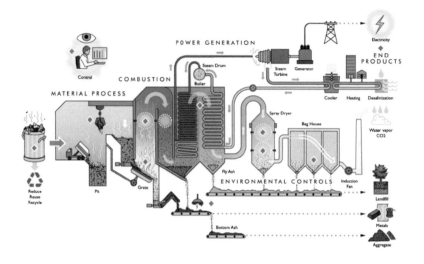

Figure 67. A schematic of a waste-to-energy plant. It not only uses any combustible waste that is not suitable for reuse or recycling, but it can recover metals from composite products such as tires with steel belting and wood with nails in it. The bottom ash can be used as aggregate. Only the fly ash and waste from the pollution control equipment require landfilling.

waste-to-energy plants that produce enough electricity for 18 million people, or four percent of the population, and enough heat for 15 million people or about 3.5 percent of the population (see Fig. 68).[32] Japan has 380 waste-to-energy plants.[33]

According to a January 2019 report from the International Energy Agency (IEA), China now has the largest installed waste-to-energy capacity of any country globally with 7.3 gigawatts across 339 plants since the end of 2017. One gigawatt is the equivalent of one large nuclear reactor or two large coal powerplants. The industry has grown by one gigawatt per year on average in the past five years, and now represents the largest form of bioenergy capacity in the country, capable of managing just over 100 million tonnes of solid waste per year.[34]

The United States, which in 2017 produced 268 million tons of solid waste, sent 53 percent to landfills, recycled or composted 35 percent,

---

32. CUWEP, "Waste-to-Energy Plants in Europe in 2017," updated to 2019.
33. Nikkei Staff Writers, "Southeast Asia's trash, Japan Inc.'s power-generating treasure," *Nikkei Asian Review*, June 23, 2019. https://asia.nikkei.com/Spotlight/Environment/Southeast-Asia-s-trash-Japan-Inc.-s-power-generating-treasure.
34. Ben Messenger, "IN DEPTH: Waste to Energy – China Drives an Asian Awakening," *Waste Management World*, June 6, 2019. https://waste-management-world.com/a/in-depth-waste-to-energy-china-drives-an-asian-awakening.

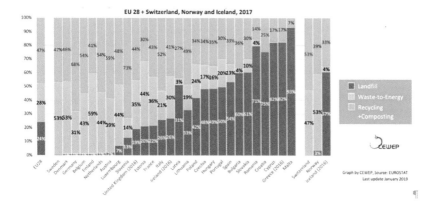

Figure 68. A comparison of the fate of municipal solid waste in Europe shows a wide divergence among most western and eastern European countries. Germany directs 59 percent of its waste to produce energy and Slovenia recycles or composts 73 percent of its waste. Many eastern European countries send a large percentage of their waste to landfills as is the case in North America.

and 13 percent was used for waste-to-energy.[35] In 2018, Canada produced 34 million tons of municipal waste of which 27 percent was recycled or composted. Statistics for waste-to-energy are difficult to find but it appears no more than three to five percent of Canada's waste stream is used for waste-to-energy. This means at least 60 percent of their waste ends up in landfills where it emits methane and often leaches contaminated water into the soil.

There are a number of reasons for this reliance on landfills in the US and Canada, including the fact that land in most of North America is less expensive than it is in more densely populated countries like Japan and Europe. But one of the main reasons there are so few waste-to-energy plants in the US and Canada is because of the strong opposition to them from the "green" movement. When one looks into this it becomes apparent that green activists are not only opposed to burning fossil fuels, they are also pretty much against the burning of wood; and they don't prefer the burning of waste that is not suitable for reuse or recycling. In other words, they are generally opposed to fire, also known as combustion. This opposition has resulted in a lack of political support for waste-to-energy options, while unreliable, expensive wind and solar energy options receive massive subsidies. Waste-to-energy plants

---

35. EPA, "National Overview: Facts and Figures on Materials, Wastes and Recycling," United States Environmental Protection Agency, 2017. https://www.epa.gov/facts-and-figures-about-materials-waste-and-recycling/national-overview-facts-and-figures-materials.

not only displace fossil fuels, they also result in much less waste, including plastic waste that is discarded into the environment.

Critics of waste-to-energy claim materials that should be recycled will end up in the furnace, but this is not the case. It can easily be seen from Figure 67 that the countries with the highest percentages of waste-to-energy recovery also have the highest rates of recycling. It is the availability of landfills, basically dumps, that reduces the recycling and waste-to-energy programs. Germany, Denmark, Sweden, and Finland have set a high bar that can be achieved in any country with the political will and technical know-how. Canada and the United States have a long way to go.

China and South Asia are now leading the movement towards more use of combustible waste for energy. Perhaps North Americans will see the wisdom of this strategy sooner than later (see Fig. 69).

Figure 69. China has recently built the world's largest waste-to-energy plant in Shenzhen. It will burn more than 5,000 tons of waste per day, producing 168 megawatts of electricity. The roof is covered with 44,000 square meters (474,000 square feet) of solar panels. The visitors center will focus on education in waste management and recycling.*

*State of Green, "World's Largest Waste to Energy Power Plant, China," undated. https://stateofgreen.com/en/partners/babcock-wilcox-volund/solutions/world-s-largest-waste-to-energy-power-plant/.

CHAPTER 7
Genetically Modified Foods Contain Something Harmful. What is it?

If there is something harmful in GMOs, it must be invisible, even under an electron microscope with 10 million times magnification.[1] So let's just make up some really scary narratives about the unforeseen damage to health and the environment that just might be caused by these unnatural mutations developed by evil seed companies. The fact is that after 25 years of genetically modified organisms of many types being grown and eaten around the world in billions of meals there has not been a single verified case of harm. But that does not seem to deter the doomsayers. They are certain there must be something very harmful in those plants, even though it is indiscernible and has no name or chemical formula.

Of all the fabricated scare stories today, this is probably the most serious one as it is costing millions of lives, especially among children and pregnant women. There is simply nothing in the genetically modified organisms that are being grown around the world today that could cause harm. This alleged danger is not only invisible, it simply doesn't exist. Yet the ignorant, and sometimes hateful, people who perpetrate this lie will not relent. There's good money for activists, for the media,

---

1. Wikipedia, "Electron Microscope," July 23, 2020. https://en.wikipedia.org/wiki/Electron_microscope.

Figure 70. The blog in which this image was used begins with these claims, "scientific studies have proven that genetic modification of foods can cause cancer and create allergies, toxins, antibiotic resistant diseases, carcinogenic, anti-nutritional, and other nutritional problems. Monsanto and our government have opened a Pandora's Box in our food supply which can create serious health hazards for all who ingest these genetically modified products."* All of these claims are false.

* Jim Bonham, "Dangers of Genetically Modified Food," Farmer Jim's Blog, July 20, 2010. http://jimbonham.com/blog/genetically-modified-foods/.

and for the leftist politicians so why would they give up on a sure thing? But unlike the climate catastrophe narrative, no scientists of any merit have joined the effort to discredit this technology that has contributed immensely to our agricultural productivity, at the same time reducing its environmental impact. And there is so much more we can accomplish with genetic science if permitted to move forward.

The propaganda began early with the first publication of the term "Frankenfood" appearing in a letter to the *New York Times* on June 6, 1992, commenting on the newly created Flavr Savr Tomato, one of the first GMO's to hit the market. It read in part:

> *If they want to sell us Frankenfood, perhaps it's time to gather the villagers, light some torches, and head to the castle.*[2]

2. Catherine Mazanek, "Frankenfoods: Conceptualizing the Anti-GMO Argument in the Anthropocene," Miami University, undated. https://journals.psu.edu/ne/article/download/60058/59862.

The terms "Frankenfoods" and "Frankenstein foods" became the battle cry for the scare campaign. The use of these terms peaked in 1999 and the damage was done.[3] Soon, other clever anti-GMO marketing terms emerged. One of the most effective denunciations of GMOs was the specter of "terminator seeds," invoking the fear of killer cyborg foods, in reference to the 1984 *Terminator* film. This phrase referred to the genetically modified food crops that had been modified to be seedless. This was met with organized outrage as Monsanto and other seed companies were accused of "addicting" farmers to their seeds, making it impossible for the farmers to use the seeds from part of their previous harvest to plant for the next year's crop. They would be forced to buy fresh seeds yearly from the seed company.

No one seemed to notice two pretty important facts. First, modern farmers almost never save seeds from their crop as they are usually not true to the improved hybrids they planted and simply don't perform as well. That is the main reason farmers usually buy new seeds each year. Second, there were already plenty of seedless food crops that had been produced by conventional breeding techniques. Examples include watermelons, bananas, tomatoes, grapes, cucumbers, oranges, lemons, and limes. No one had ever thought to condemn seed companies for doing this, yet millions marched in the streets to ban "Frankenfoods" and "terminator seeds."[4]

The underlying philosophy was that genetically modified organisms amounted to mere mortals "playing God" and this activity must not be permitted. By this standard virtually everything we've done since harnessing fire and using stone tools is playing God, never mind supersonic aircraft and nuclear reactors. This "humans playing God" accusation was underscored by the assertion that the "precautionary principle" must be invoked just in case something bad arises. And of course, there was the fear that multinational seed companies would become monopolies and "control the global food supply and dominate the world."[5] It was never explained how selling superior seeds to farmers, who are free to buy whatever seeds they prefer, could bring about a global dictatorship.

---

3. Lina Hellsten, "Focus on Metaphors: the Case of "Frankenfood" on the Web," *Journal of Computer-Mediated Communication*, July 1, 2003. https://academic.oup.com/jcmc/article/8/4/JCMC841/4584272.
4. *The Guardian*, "Millions march against GM crops," Associated Press, May 26, 2013. https://www.theguardian.com/environment/2013/may/26/millions-march-against-monsanto.
5. Ariel Poliandri, "Considering the three types of reasons people oppose GMO foods," Genetic Literacy Project, January 9, 2015. https://geneticliteracyproject.org/2015/01/09/considering-the-three-types-of-reasons-people-oppose-gmo-foods/.

Every individual of every species that is produced through sexual reproduction is genetically modified. None of us are genetically identical to our parents, that is, each of us are "modified" due to the unique random combination of our parent's genes. Identical twins are genetically identical to each other but not to their parents. This point is countered by the fact that GMOs are the result of moving genes from one species to another, which is "unnatural." Not so: "gene flow" between species has been occurring pretty much since life began and is a major factor in evolution. But, unlike the highly controlled recombinant DNA technique used to produce genetically modified foods, natural gene flow is entirely random, and often produces negative results.[6]

Ever since bacteria evolved about 3.5 billion years ago, they have been transporting DNA from one organism to another. This abstract from *ScienceDirect* on the subject states this clearly.

> *Horizontal gene transfer, which means gene transfer without reproduction, is widespread among Bacteria, Archaea, and unicellular eukaryotes, and also occurs in higher eukaryotes. Genetic material can be transferred horizontally between evolutionary distant organisms. Different species possess different mechanisms for horizontal transfer. The ubiquity of horizontal gene transfer makes it a major evolutionary phenomenon. Understanding the widespread of horizontal gene transfer has shaken the classic treelike view (of) evolution.*[7]

Eukaryotes are organisms whose cells contain organelles such as the nucleus, chloroplast, and mitochondria. Humans are a higher eukaryote and bacteria can transport genes from other species into our bodies too.

Mallard ducks often breed with other species of ducks including black ducks and Muscovy ducks to produce hybrid species. There are more than 400 known waterfowl hybrids that have resulted from crossbreeding among different species.[8] So, there is nothing unnatural about gene exchange among species and GMOs are simply another example of this phenomenon.

Few people are aware that during the past century many of our food crop varieties have been produced by both processes of bombarding

---

6. "Horizontal Gene Transfer," Wikipedia, August 12, 2020. https://en.wikipedia.org/wiki/Horizontal_gene_transfer.
7. N. Yutin, "Horizontal Gene Transfer," *ScienceDirect*, Brenner's Encyclopedia of Genetics, 2013, pp530-532. https://www.sciencedirect.com/science/article/pii/B978012374984000735X?via%3Dihub.
8. Jennifer Kross, "Hybrid Waterfowl," Duck Unlimited, undated. https://www.ducks.org/conservation/waterfowl-research-science/waterfowl-hybrids.

seeds with radiation[9] and by soaking them in mutagenic chemicals.[10] Many of the varieties used in "organic farming" have been developed with these techniques.

This excerpt is from an excellent article in *The Conversation* written by a Postdoctoral Scholar in plant biology at the University of California, Berkeley:

> *Mutation breeding, which in my opinion is also a type of biotechnology, is already used in organic food production. In mutation breeding, radiation or chemicals are used to randomly make mutations in the DNA of hundreds or thousands of seeds which are then grown in the field. Breeders scan fields for plants with a desired trait such as disease resistance or increased yield. Thousands of new crop varieties have been created and commercialized through this process, including everything from varieties of quinoa to varieties of grapefruit. Mutation breeding is considered a traditional breeding technique, and thus is not an 'excluded method' for organic farming in the United States.[11]*

Both radiation mutation and chemical mutation breeding are entirely unpredictable. Most of the results are either benign or negative and are discarded. But the occasional positive results are often adopted as new superior strains and become the most desired seeds for both conventional and organic farming. This is very similar to the way evolution and natural selection operate. Most natural mutations are either benign or negative and do not survive in the long run. But the occasional mutation that confers an advantage to the species survives and often displaces previous traits that were not as advantageous. Thus, the species adapts to remain viable and competitive in a very competitive global ecosystem.

By contrast both genetic engineering and gene editing are much more precise interventions. These two Wikipedia entries provide a thorough explanation of the various techniques and many references for the serious student.[12,13] The scientists and agronomists using these

---

9. H. Brunner, "Radiation induced mutations for plant selection," *ScienceDirect*, June-July 1995, pp589-594. https://www.sciencedirect.com/science/article/abs/pii/0969804395000968.
10. Joanna Jankowicz Cieslak, et al., "Chemical Mutagenesis of Seed and Vegetatively Propagated Plants Using EMS," Current Protocols in Plant Biology, December 1, 2016. https://currentprotocols.onlinelibrary.wiley.com/doi/10.1002/cppb.20040.
11. Rebecca Mackelprang, "Organic farming with gene editing: An oxymoron or a tool for sustainable agriculture?" *The Conversation*, October 10, 2018. https://theconversation.com/organic-farming-with-gene-editing-an-oxymoron-or-a-tool-for-sustainable-agriculture-101585.
12. Wikipedia, "Genome Editing," August 24, 2020. https://en.wikipedia.org/wiki/Genome_editing.
13. Wikipedia, "CRISPR gene editing," August 26, 2020. https://en.wikipedia.org/wiki/CRISPR_gene_editing.

techniques know in advance what they are trying to accomplish by inserting new genes, removing genes, or by turning genes on or off. Just the same there is a requirement with all these genetic techniques to demonstrate their safety before they are released to the market. The onerous nature of these requirements, much of which is due to the anti-GMO movement, has resulted in a regulatory regime that often requires more than $100 million to gain approval for each new variety of seed.

The champions of the grass-roots organic back-to-the-land philosophy have created a regime where only large agri-business companies can afford to advance the genetic improvement of our food and fiber crops using biotechnology. Once the regulatory hurdles are completed the large companies then have the best seed, and farmers will buy them over the now obsolete seeds the smaller companies are selling. Ironically, the anti-GMO movement has helped create the very outcome they warned against in the first place, an advantage for the larger seed companies over the smaller ones.

Virtually every major science, medical, and nutritional organization has declared that genetically modified foods are as safe or safer than conventional foods. They are safer because they are so much more rigorously tested than new varieties of conventional crops.

In 2010 the European Commission of the European Union declared:

The main conclusion to be drawn from the efforts of more than 130 research projects, covering a period of more than 25 years of research and involving more than 500 independent research groups, is that biotechnology, and in particular GMOs, are not per se more risky than e.g. conventional plant breeding technologies.[14]

As early as 2004 the US National Academy of Sciences published this statement, thus confirming that GMOs are safer than conventional foods:

In contrast to adverse health effects that have been associated with some traditional food production methods, similar serious health effects have not been identified as a result of genetic engineering techniques used in food production. This may be because developers of bioengineered organisms perform extensive compositional analyses to determine that each

---

14. "A decade of EU-funded GMO research (2001–2010)," European Commission, 2010. https://ec.europa.eu/research/biosociety/pdf/a_decade_of_eu-funded_gmo_research.pdf.

phenotype is desirable and to ensure that unintended changes have not occurred in key components of food.[15]

There is some irony in activists falsely claiming there is an "overwhelming consensus" among scientists that climate change is an emergency, whereas where there truly is an overwhelming consensus among scientists that GMOs are safe to eat they simply ignore it and spread fear on the basis of inaccurate claims.

Vitamin A Deficiency and Golden Rice

Vitamin A is an essential nutrient for all animals including humans, but animals cannot produce vitamin A themselves. They can, however, synthesize vitamin A from beta-carotene which is produced in all green plants. Beta-carotene is present in the leaves of all green plants, but its color is overpowered by the green of chlorophyll. The color of beta-carotene, ranging from yellow to orange, can be seen in many root vegetables. In fact, carrots were the inspiration for the name beta-carotene due to their bold expression of its color. Many of our staple foods contain adequate beta-carotene to satisfy our daily needs. These include sweet potatoes ("yam-fries" are actually sweet potatoes), carrots, dark leafy greens like spinach, butternut squash, cantaloupes, romaine lettuce, and more.[16]

By far, the deadliest lack of essential nutrients in humans is a vitamin A deficiency. Many of our staple foods contain beta-carotene but rice does not. Rice is the staple food for most people in the tropics and the poorest of those people cannot afford much more than a cup of rice each day to keep themselves alive. Because of this about 250 million pre-school children are deficient in vitamin A (see Fig. 71, page 136). In addition, pregnant women, who need vitamin A for both themselves and their unborn child, require more of the nutrient than other people and are subject to this deficiency. Each year between one to two million people, mostly children, die from a vitamin A deficiency. Despite charitable programs to deliver vitamin A tablets to underdeveloped places, millions of people still don't get them, and the death toll continues.[17]

---

15. National Research Council, et al., "Safety of Genetically Engineered Foods," *The National Academies Press*, 2004. https://www.nap.edu/catalog/10977/safety-of-genetically-engineered-foods-approaches-to-assessing-unintended-health.
16. Daisy Whitbread, "Top 10 Foods Highest in Beta Carotene," My Food Data, August 6, 2020.
17. "Mission 2014: Feeding the World – Micronutrient Deficiency," Massachusetts Institute of Technology, 2014. http://12.000.scripts.mit.edu/mission2014/solutions/micronutrient-supplementation.

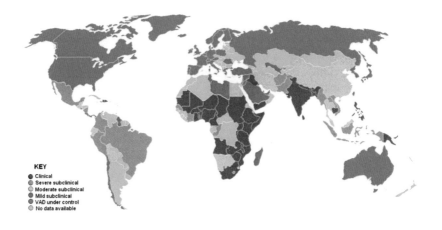

Figure 71. While Vitamin A deficiency is highest in Africa and Asia it is also a serious problem in Mexico. This is due to a number of factors, including inadvertently breeding the beta-carotene out of yellow corn to make white corn for tortillas, eating only the beans and not the green pods of bean plants, and eating white rice which has no beta-carotene at all.

In addition, between 250,000 to 500,000 vitamin-A-deficient children become blind every year, with half of them dying within 12 months of losing their sight.[18]

It was the scientists working for the United Nations Children's Fund (UNICEF) that first discovered that Vitamin A was not only necessary for eyesight but was also key to a functioning immune system. This led to the realization that many children who were dying from malaria, dengue, diarrhea, etc., would have recovered if not for their deficiency in vitamin A, which was compromising their immune systems.[19] Once this was understood, efforts were made to breed a strain of rice that would incorporate beta-carotene into the kernels. Rice does have beta-carotene in its leaves, as do all green plants, but it soon became clear that conventional breeding would not succeed in producing grains of rice with beta-carotene in them.

The 1980s and 1990s saw the birth of recombinant DNA technology – soon to be known as genetic engineering – that produced genetically modified organisms by moving genes from one species to another

---

18. "Micronutrient Deficiencies – Vitamin A Deficiency," World Health Organization. https://www.who.int/nutrition/topics/vad/en/.
19. Melissa Miller, et al., "Why Do Children Become Vitamin A Deficient?" *The Journal of Nutrition*, September 1, 2002. https://academic.oup.com/jn/article/132/9/2867S/4687677.

in order to confer a desirable trait in that species. To date most of the novel GMOs have been produced to resist insect pests and/or to tolerate herbicides so that the crop can grow while herbicides are used to control weeds. Other advances include salmon that grow through the winter and therefore reach market size far sooner than wild salmon, as well as papayas and squash that are resistant to viruses.

In 1999 two European scientists, Dr. Ingo Potrykus, a professor at the Swiss Federal Institute of Technology in Zurich, and Dr. Peter Beyer of the University of Freiburg, announced that they had produced a rice plant that contained beta-carotene in the actual rice grains, which gave it a distinctive yellow color. They created the Golden Rice Project with the intention of taking the new Vitamin-A-rich rice to market and bringing an end to one of the worst humanitarian crises in the world (see Fig. 72). The July 31, 2000 edition of *Time* magazine carried a photo of Ingo Potrykus with the caption: *"This Rice Could Save a Million Kids a Year."* But the ominous subhead said, *"...but protesters believe such genetically modified foods are bad for us and our planet. Here's why."*

Figure 72. Dr. Ingo Potrykus pictured with a Golden Rice plant on the cover of *Time* magazine. He and his colleagues have worked for more than 20 years to get Golden Rice approved.*
* J. Madeleine Nash, "This Rice Could Save a Million Kids a Year," *Time* magazine, July 31, 2000 http://content.time.com/time/magazine/article/0,9171,997586-1,00.html.

Notice it doesn't say "scientists believe" but rather "protesters." Thus began a struggle for the truth and public opinion that continues today. It is one of the saddest stories of this era and it is caused by ignorance and greed.

It took four years to negotiate the details of patent rights and to help the countries that needed Golden Rice the most to adopt the necessary regulatory frameworks under international law. In 2004, the first field trial was harvested in Louisiana and by the end of that year, with assistance from Syngenta, the potency of the beta-carotene had been multiplied by 23 times over the 1999 prototype.[20] All the while an avalanche of opposition was growing among the anti-GMO movement. Greenpeace International stated that:

> *Not only is golden rice an ineffective tool to combat VAD (vitamin A deficiency) it is also environmentally irresponsible, poses risks to human health, and compromises food security.*[21]

Of course, there were zero particulars expanding on the claims of it being "an ineffective tool," "environmentally irresponsible," "a risk to human health," or the ridiculous assertion that rice with beta-carotene could compromise "food security." It was simply fearmongering and dishonesty from people who are generally ignorant in scientific matters and of genetic science in particular. But it sells newspapers and brings in donations.

Progress continued and the project was taken up by the International Rice Research Institute in the Philippines and their national counterparts in Bangladesh and Indonesia, countries among those with the most severe cases of Vitamin A deficiency. Financial and organizational support was provided by the Rockefeller Foundation, the Golden Rice Humanitarian Board, the Bill and Melinda Gates Foundation, Helen Keller International, and many others. By 2012, large-scale field trials were underway with the aim of genetically modifying the rice varieties preferred in each country, a time-consuming undertaking (see Fig. 73).

On August 8, 2013 the Associated Press, quoting a Greenpeace spokesperson, reported that "Filipino farmers" had destroyed a field trial of Golden Rice at the International Rice Research Institute

---

20. "History of the Golden Rice Project – The Road to the Farm is Bumpy." http://www.goldenrice.org/Content1-Who/who2_history.php.
21. Janet Cotter, "Golden Illusion – The Broken Promises of 'Golden' Rice," Greenpeace International, October 2013. https://storage.googleapis.com/planet4-international-stateless/2013/10/08786be5-458-golden-illusion-ge-goldenrice.pdf.

Figure 73. Field trials of Golden Rice at the International Rice Research Institute in the Philippines. Many local varieties of rice are preferred depending on weather, latitude, soil type, culture, cuisine, etc. The technique of using genetic modification to produce a new variety is a very complex and time-consuming process.

because they feared it would "contaminate" their rice fields. The original Associated Press release is no longer on the web but reports from *New Scientist* and the BBC confirm that local farmers were blamed by the leftist urban activists who were the ones that actually destroyed the field trial.[22,23]

Environmental activist Mark Lynas, who had formerly strongly opposed genetically modified crops but had changed his mind, traveled to the Philippines and discovered the truth. The truth was that Greenpeace bussed in mainly young urban activists for the explicit purpose of destroying the field trials (see Fig. 74, page 140).[24]

A few hours after this event, my wife Eileen, my late brother Michael, and I were sitting at our kitchen table having coffee in our beach house – the house Eileen and I had built by hand in 1974 – in

---

22. Michael Slezak, "Militant Filipino farmers destroy Golden Rice GM crop," *New Scientist*, August 9, 2013. https://www.newscientist.com/article/dn24021-militant-filipino-farmers-destroy-golden-rice-gm-crop/?ignored=irrelevant.
23. Matt McGrath, "'Golden rice' GM trial vandalised in the Philippines," BBC, August 9, 2013. https://www.bbc.com/news/science-environment-23632042.
24. Mark Lynas, "The True Story About Who Destroyed a Genetically Modified Rice Crop," *Slate*, August 26, 2013. https://slate.com/technology/2013/08/golden-rice-attack-in-philippines-anti-gmo-activists-lie-about-protest-and-safety.html.

Figure 74. Leftist urban activists bussed into the International Rice Research Institute's field trials of Golden Rice in order to destroy them. Inset: Some of the 100 or so activists ripping the Golden Rice plants out by the roots. These are not farmers.

Winter Harbor. The Associated Press report was on my laptop and I already figured Greenpeace was lying about who actually carried out the deed. As a frequent speaker at conferences amongst professionals, I had been talking about Golden Rice for some years now, as well as the crimes against humanity being committed by its enemies. Right then and there we vowed to get up and get active. The Allow Golden Rice Now Campaign began at that kitchen table.

By this time there were hundreds of anti-GMO organizations around the world campaigning every day. Not one organization was campaigning for Golden Rice. Many people were involved at the scientific level and in negotiating the complicated patent rights to make sure Golden Rice was not monopolized but instead, it was given freely to farmers who wished to grow and sell it in local markets. To make matters worse, the anti-GMO groups saw Golden Rice as a real threat to their campaigns because it had the potential to be a major humanitarian success. These people did not want a GMO-success story ruining their crusade of misinformation and outright lies. We were taking on a real juggernaut that had hundreds of millions of dollars at its disposal. But we had the truth on our side.

I had recently met businessman Cary Pinkowski, and was giving him advice on the environmental aspects of a gold mine he was hoping to develop in northern Spain. Cary and his wife Katya offered to host

Figure 75. The first time anyone campaigned in public for Golden Rice was October 11, 2013. Greenpeace has been the leader in opposing Golden Rice from the start. From left to right: myself, our son Jonathan, my wife Eileen, our long-time friend Sheila Patterson, Jonathan's wife Sarah, and on the other end of the banner, my late brother Michael's son Ariel.

a fundraiser for our campaign in downtown Vancouver and invited a good number of their friends and business associates. The crowd listened to our story about Golden Rice and our plans to travel to Europe and Asia to protest against Greenpeace at their offices, as well as hold media conferences to get the word out. We then received enough support to get the campaign underway. We launched a website that provides the history and mission of our campaign that reached more than 30 million people with a positive message about Golden Rice. The Media section of the website documents our success in Europe, North America, and Asia in reaching the public.[25]

Our first opportunity to protest against Greenpeace fell right into our laps when the organization announced they would visit Vancouver in their $32-million-motor vessel, the Rainbow Warrior III.[26] They call it a sailing ship, but it has an 1,800 HP diesel engine, so it should be more correctly labeled a "sail-assist." Nothing wrong with that except they never say the word "engine." On the 11th and 12th of October in 2013, we picketed Greenpeace beside the Rainbow Warrior III where it was docked in North Vancouver (see Fig. 75). The fact that one of the

---

25. Michael Moore, et al., "Allow Golden Rice Now," Allow Golden Rice Society, 2013-2016. http://allowgoldenricenow.org/wordpress/help-the-campaign/.
26. Greenpeace Danmark, "Virtual Tour of Rainbow Warrior III," November 3, 2011, https://www.youtube.com/watch?v=6P2q-QD545c.

co-founders of Greenpeace was protesting against them for humanitarian reasons caught the media's interest and the campaign had begun.[27,28]

Thus began three years of intensive campaigning. We focused on Europe where the center of opposition to genetic modification technology can be found, beginning with the headquarters of Greenpeace Germany in Hamburg. They had become the most powerful of all the Greenpeace national organizations and were at the center of the anti-Golden Rice campaign. There we were joined by Horst Rehberger and Uwe Schrader of the Free Democratic Party. Uwe had been the head of a genetic modification educational farm before his staff were violently attacked in the night by anti-GM thugs who destroyed the facility. This marked the beginning of the end for genetically modified crops in nearly all of Europe. Horst had been Finance Minister in the former East German state of Saxony-Anhalt after reunification. Both were strong supporters of agricultural biotechnology and of Golden Rice in particular. Both had been introduced to me by my friend Klaus Amman, at that time Director of the Botanic Garden in Bern, Switzerland, who had also joined the campaign.

During three trips to Europe in 2013 and 2014 we visited eight key countries where Greenpeace had national offices and demonstrated at them with sufficient media attention. We even gained some support from left-leaning media as they could see the humanitarian aspect of saving the lives of millions of children might outweigh the non-existent supposedly bad thing in Golden Rice.

In 2015 we traveled to South Asia where we were joined by Hans-Joerg Jacobsen and Paul Evans. We visited the International Rice Research Institute in the Philippines, their counterpart in Bangladesh, and in India where Golden Rice would be beneficial for hundreds of millions of people (see Fig. 76). All these visits, media coverage, protests, and meetings are documented in detail on the campaign website.[29]

Finally, in December of 2019, 20 years after it was developed and 15 years after it was ready to be grown, the government of the Philippines

---

27. Patrick Moore, "By opposing Golden Rice, Greenpeace defies its own values – and harms children," *The Globe and Mail*. October 8, 2013. https://www.theglobeandmail.com/opinion/by-opposing-golden-rice-greenpeace-defies-its-own-values-and-harms-children/article14742332/.
28. Hank Campbell, "Co-Founder of Greenpeace: Greenpeace is Wrong About Golden Rice," Science 2.0, October 21, 2013. https://www.science20.com/science_20/cofounder_of_greenpeace_greenpeace_is_wrong_about_golden_rice-122754.
29. Michael Moore, et al., "Allow Golden Rice Now," Allow Golden Rice Society, 2013-2016. http://allowgoldenricenow.org/wordpress/help-the-campaign/.

Figure 76. Meeting with Bangladesh Minister of Agriculture, Matia Chowdhury, a strong advocate for agricultural biotechnology in a country that is highly dependent on farming. From the left, Uwe Schrader, two ministry staff, Eileen Moore, myself, Minister Chowdhury, Horst Rehberger, Hans-Joerg Jacobsen, and ministry staff.

announced that it had approved Golden Rice for both human consumption and animal feed.[30]

Just one month later, Greenpeace Philippines filed a petition with the government of the Philippines to revoke their approval.[31] This will not likely be achieved, given the thorough and lengthy review done by the authorities who found not one shred of evidence that Golden Rice could cause harm to people or to the environment. Adrian Dubock, project manager for the Golden Rice Project, wrote a particularly good rebuttal in response to Greenpeace's petition for the Genetic Literacy Project.[32]

---

30. International Rice Research Institute, "Philippines approves Golden Rice for direct use as food and feed, or for processing," December 18, 2019. https://www.irri.org/news-and-events/news/philippines-approves-golden-rice-direct-use-food-and-feed-or-processing.
31. Greenpeace Philippines, "Greenpeace submits formal appeal to Deptartment of Agriculture to revoke 'golden rice' approval," December 24, 2019. https://www.greenpeace.org/philippines/press/4025/greenpeace-appeal-department-of-agriculture-to-revoke-golden-rice-approval/.
32. Adrian Dubock, "Viewpoint: "On the wrong side of humanity and science", Greenpeace Philippines launches last gasp effort to derail GMO Golden Rice approval," Genetic Literacy Project, January 29, 2020. https://geneticliteracyproject.org/2020/01/29/viewpoint-on-the-wrong-side-of-humanity-and-science-greenpeace-philippines-launches-last-gasp-effort-to-derail-gmo-golden-rice-approval/.

Hopefully other countries where Golden Rice could save countless lives will follow the Philippines' lead and end the suffering and grief experienced by so many families for so many years. While the "green" left calls for socialism and redistribution of wealth, they are fighting against a cure for the very poorest people in the world; when one cup of Golden Rice per day could save millions of them from blindness and eventual death. They die quietly and unnoticed.

CHAPTER 8
Fear of Invisible Radiation from Nuclear Energy

Unlike the non-existent, fabricated harmful element claimed to reside in genetically modified organisms, radiation is real, and most of it is actually invisible. There are many kinds of radiation ranging from very longwave to very short wave. The full range of radiation of all wavelengths is called the electromagnetic spectrum (see Fig. 77, page 146). These bands of radiation, like visible light, have no mass. The waves are formed by pure energy called photons.

Solar radiation is composed not only of non-ionizing visible light, but also contains ultraviolet (UV) radiation, which is ionizing, invisible, and a serious health risk. Ultraviolet radiation is from the life-giving Sun, can cause sunburns, and if one is very over-exposed it can even cause skin cancers such as melanoma. In other words, the Sun's radiation is actually classified as a carcinogen.[1] But we are not going to ban sunlight, however, because it has proven positive effects for the health of most species on Earth, including humans. Obviously, solar radiation is necessary for photosynthesis in green plants, and without which life

---

1. Sid Kirchheimer, "UV Radiation Listed as Known Carcinogen," WebMD, December 30, 2002. https://www.webmd.com/cancer/news/20021230/uv-radiation-listed-as-known-carcinogen#1.

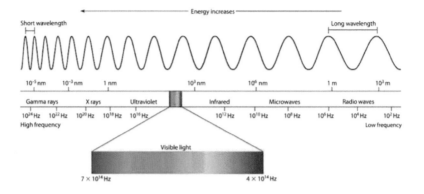

Figure 77. This is the full range of radiation in the electromagnetic spectrum, ranging from low-energy long wavelength radio waves to high-energy short wavelength gamma rays. Visible light is a relatively small part of the electromagnetic spectrum. To the right of visible light radiation is called non-ionizing, meaning it does not harm life's basic functions, whereas ionizing radiation to the left of visible light has enough energy to damage living tissue and DNA.*

* Environmental Protection Agency, "Radiation Basics," June 11, 2019. https://www.epa.gov/radiation/radiation-basics.

would be reduced to a pitiful few species of bacteria. In truth, without the Sun there would be no life. Animals like us benefit from sunlight because it increases vitamin D production, in turn boosting the immune system and increasing the absorption of essential nutrients like calcium and phosphorous.[2]

It is unfortunately all too common that some substances or types of radiation are simplistically labeled as "toxic." Actually, there are many substances and radiation types that are beneficial to receive at low levels, while they can be toxic at higher levels. In fact, the first law of toxicology is "the dose makes the poison."[3] This is as true for sunlight as it is for table salt – which is an essential nutrient when consumed in moderation, but swallowing four to five tablespoons of it can be fatal.[4]

---

2. Healthline, "The Benefits of Vitamin D," November 13, 2017. https://www.healthline.com/health/food-nutrition/benefits-vitamin-d#how-much-you-need.
3. "The Dose Makes the Poison," ChemicalSafetyFacts.org, 2020. https://www.chemicalsafetyfacts.org/dose-makes-poison-gallery/.
4. Norm R. C. Campbell, "A Systematic Review of Fatalities Related to Acute Ingestion of Salt. A Need for Warning Labels?" Nutrients, June 23, 2017. https://www.ncbi.nlm.nih.gov/pmc/articles/PMC5537768/#:~:text=The%20lethal%20dose%20was%20estimated,be%20discerned%20from%20our%20review.

You have probably noticed that nothing has yet been said about "nuclear radiation" which is the subject of this chapter. This is a very technical and information-rich subject. One could spend a lifetime studying it and still not attain all there is to know. Here goes.

There are three different types of radiation from radioactive elements: alpha, beta, and gamma radiation; each being named after the first three letters in the Greek alphabet. All three types of radiation are invisible. Whereas all the radiation in the electromagnetic spectrum is pure energy, two of these types of radiation that emanate from radioactive elements consist of particles that have mass.[5]

Alpha radiation is a particle that has two protons and two neutrons and is therefore identical to the nucleus of a helium atom. Beta radiation is composed of either an electron or positron and is miniscule compared to an alpha particle. Alpha radiation poses little risk if it is external to the body as it is blocked by clothing and skin. But beta radiation is more dangerous as it can penetrate the skin and cause burns. Both alpha and beta radiation can be of significant risk if they are ingested in a quantity capable of overwhelming the body's ability to repair the damage they cause.

Gamma radiation is a very strong ionizing radiation that can penetrate right through the human body, like X-rays, and cause a level of harm, relative to the dose. It is a fact that, as with UV radiation from the Sun, all three types of radiation from radioactive elements such as uranium, radium, and radon have the potential to cause harm. But it's also a fact that we have ways of protecting ourselves from them by shielding with concrete and lead, and with personal protective equipment worn by employees working near all nuclear reactors.[6] The average nuclear worker in the United States receives 150 millirems (mrem) of radiation exposure per year. By contrast just living in the US results in an average exposure per person of 620 mrem. Bananas, granite, Brazil nuts, and smoke detectors emit radiation. No negative health effects have ever been detected from exposures below 10,000 mrem.

Nuclear energy is one of the safest, if not the safest technology, for generating electricity on the basis of casualties per unit of energy produced. All the same, as a serious student of science during the late 1960s and early 1970s, even I came to fear nuclear energy due to the powerful propaganda campaign against it. It was not until the 1980s,

---

5. "What Are The Different Types of Radiation?" US Nuclear Regulatory Commission, October 19, 2018. https://www.nrc.gov/reading-rm/basic-ref/students/science-101/what-are-different-types-of-radiation.html.
6. "Radiation Protection for Nuclear Employees," Duke Energy, August 21, 2012. https://nuclear.duke-energy.com/2012/08/21/radiation-protection-for-nuclear-employees.

after I left Greenpeace, that I re-educated myself about nuclear energy and realized the truth about this fascinating invention. I realized that nuclear energy had been unfairly lumped in with nuclear weapons as something evil. But it became clear upon further study that nuclear energy should be considered in the same regard as nuclear medicine – a beneficial use of nuclear technology.

When fossil fuels become scarce, even if it's not until centuries from now, nuclear will likely be the source of the majority of our energy for millennia into the future. There is fuel available for many thousands of years. I'm not saying the day will not come when we must face up to a global energy crisis; it's just not any time soon, unless it is a shortage of our own making.

The "Linear-No-Threshold" Hypothesis of Nuclear Radiation Effects

Unlike the widely accepted premise of UV solar radiation effects – which can be very harmful at high exposures but is actually beneficial at low exposures – the rules for radiation exposure, specifically for nuclear energy plant workers, are based on an entirely different standard. It is called the "linear-no-threshold" hypothesis and it assumes any exposure to radiation, beginning at zero, is harmful – and twice as much is twice as harmful. This means that nuclear radiation exposure must be strictly minimized, rather than allowing an amount of exposure that is under the threshold of what is considered a harmful amount. This makes nuclear energy much more expensive than it needs be.

The linear-no-threshold standard is patently false, but it has been forced on the industry by the anti-nuclear movement as one means of reducing the cost-effectiveness of the technology. Only China, France, and Japan have rejected the linear-no-threshold standard, and rightly so. The fact is all organisms, including ourselves, have a cellular repair mechanism that is capable of repairing damage to our bodies while it is occurring. So long as our bodies are able to repair themselves faster than they are being damaged, there is no net damage. This exposes the lie concerning the "no threshold" rule. There is clearly a threshold below which there is no net damage. This is true with all toxic substances and solar UV radiation. Until this is corrected, nuclear energy will continue to be unfairly penalized.

Hormesis

Hormesis is a word most people have never read or heard spoken. Hormesis (hore-mee-sis) is the theory that low levels of many agents are actually beneficial, while higher levels are harmful. The case of the essential nutrient, salt, presented above is a perfect example of hormesis. This is also the case for every essential nutrient. It is possible to consume too much of anything that is essential, to the point where it becomes harmful, including water. Here is a clear definition of hormesis:

> *Hormesis is a term used by toxicologists to refer to a biphasic dose response to an environmental agent characterized by a low dose stimulation or beneficial effect and a high dose inhibitory or toxic effect. In the fields of biology and medicine hormesis is defined as an adaptive response of cells and organisms to a moderate (usually intermittent) stress.*[7]

Many scientists believe low doses of nuclear radiation are actually beneficial, as with low levels of UV radiation, but through different mechanisms (see Fig. 78). Low levels of nuclear radiation appear to challenge the cellular repair mechanism and thus make it stronger, analogous to a vaccine teaching the immune system to be prepared for an

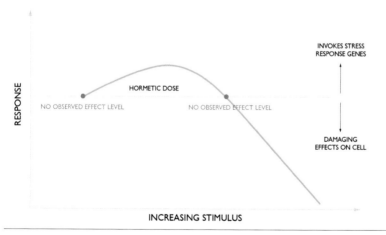

Figure 78. Key attributes of hormesis include no effect at extremely low levels, a beneficial effect with an optimum benefit level, and a damaging effect beyond which damage increases with dose.

---

7. Mark P. Mattson, "Hormesis Defined," PubMed, December 5, 2007. https://www.ncbi.nlm.nih.gov/pmc/articles/PMC2248601/.

infection. Many radiologists, whose specialty is radiation, subscribe to the reality of radiation hormesis, although you would rarely if ever hear it from the news media outlets.[8] (The paper on hormesis referenced below is particularly worth a read.)

We know that natural radiation in the environment has been a big factor in the evolution of life. Evolution has depended on random mutations caused by chemical contamination of food, water, and air and by exposure to alpha, beta, and gamma radiation. The vast majority of mutations are either harmless or negative, but the small number that are beneficial to survival are a large part of evolutionary adaptation. It's probably a bit far reaching to call all these chemicals and radiation "essential for life" but without them evolution would have been much slower, and humans would have likely never existed.

Three Mile Island, Chernobyl, and Fukushima

These three nuclear reactor accidents spread fear across whole countries and around the world. Except for Chernobyl, they were not worth nearly that much anxiety. Even Chernobyl produced a public reaction far beyond reason. More people died in the forest fire in Paradise, California than from the Chernobyl disaster, which was the only accident to cause fatalities at a civilian reactor in the history of nuclear energy. Let's briefly look at each one of these accidents.

Three Mile Island

The incident at Three Mile Island on March 26, 1979 near Harrisburg, Pennsylvania did not harm a single person and moreover, no one received a dose of radiation above the background level in the environment. Here is the synopsis of events reported by the World Nuclear Association:

> *In 1979 at Three Mile Island nuclear power plant in USA, a cooling malfunction caused part of the core to melt in the #2 reactor. The TMI-2 reactor was destroyed.*
>
> *Some radioactive gas was released a couple of days after the accident, but not enough to cause any dose above background levels to local residents.*

---

8. T. D. Luckey, "Radiation Hormesis: The Good, the Bad, and the Ugly," PubMed, September 7, 2006. https://pubmed.ncbi.nlm.nih.gov/18648595/.

> There were no injuries or adverse health effects from the Three Mile Island accident.[9]

It was unfortunate that the incident occurred shortly after the film *The China Syndrome* was released, which presented a frightening plot with a potential reactor meltdown, the core of which would melt right through the Earth and come out in China.[10] This film's concept is impossible, for even if the core melted through to the center of the Earth, gravity would make sure it remained safely right there rather than coming out in China. Of course, the movie involved a government coverup which caused a reaction of panic during the real Three Mile Island reactor failure. In the end it was just an expensive accident and important lessons were learned.

Chernobyl

The Chernobyl accident in Ukraine on April 26, 1986 really was a disaster, and not only during the incident but for many years afterwards as well. By far the highest exposure to radiation was among the workers at the plant and the emergency crew that worked to put the fire out that continued to burn for ten days. Two of the people who were in the reactor died from the blast. Among the 134 people who were diagnosed with Acute Radiation Sickness, 29 eventually died while the others recovered. A few more died some years later but the ultimate cause cannot be definitely confirmed. Of the 4,000 children who contracted thyroid cancer after the accident, 15 died, mainly due to late diagnosis.[11]

One of the worst effects of the accident befell the approximately 340,000 people who were evacuated and displaced to large tenement buildings in the outskirts of Kiev. It was there that the occurrences of suicide, drug and alcohol addiction, violence, marital collapses, and mental illness and trauma that resulted from living in these crowded urban quarters clearly outweighed the possible effects of the increased radiation exposure they would have experienced had they been left in their rural homes. The main limitation would have been importing all

---

9. "Three Mile Island Accident," World Nuclear Association, March 2020. https://www.world-nuclear.org/information-library/safety-and-security/safety-of-plants/three-mile-island-accident.aspx.
10. "The China Syndrome," Wikipedia, August 31, 2020. https://en.wikipedia.org/wiki/The_China_Syndrome.
11. Burton Bennett, et al., "Health Effects of the Chernobyl Accident and Special Health Care Programs," World Health Organization, Geneva, 2006. https://apps.who.int/iris/rest/bitstreams/51546/retrieve.

their food from areas free of contamination, a situation that occurred in their Kiev lodgings as well.

Despite many fire worker's exposure to high levels of radiation the number of fatalities that were attributed to radiation were surprisingly small. 20 years after the accident the World Health Organization concluded:

> *Epidemiological studies of residents of areas contaminated with radionuclides in Belarus, Russia, and Ukraine performed since 1986, so far have not revealed any strong evidence for radiation-induced increase in general population mortality, and in particular, for fatalities caused by leukemia, solid cancers (other than thyroid), and non-cancer diseases.*[12]

The only good thing about the Chernobyl accident is that this class of reactor, RBMK, will never be built again. The accident in 1986 was not from a melt-down event as was the case for both Three Mile Island and Fukushima. The Chernobyl explosion was an atomic explosion caused by a runaway chain reaction. This was partly due to the design having a "positive void coefficient" unlike any other commercial reactor design.[13]

During the Cold War the former Soviet Union took a short cut by modifying the reactors it was using to produce plutonium for nuclear weapons and then using them as power reactors to generate electricity. The accident at Chernobyl was not caused by routine operations but by an experiment being conducted by outside engineers while the reactor was under power. The reactor operators and the engineers failed to communicate clearly with one another and made errors in judgement. This resulted in the reactor running out of control and after eight seconds it exploded, sending a plume of radiation into the atmosphere (see Fig. 79). It took ten days to put the fire out because it was fueled by the 2,000 tons of graphite (carbon) used as a moderator in the core of the reactor.[14]

There were three other RBMK reactors at the same site as the reactor that was destroyed. They continued to operate for some years after the accident but the last one was decommissioned in 2001. There

---

12. Ibid.
13. "RBMK Reactors – Appendix to Nuclear Power Reactors," World Nuclear Association, July 2019. https://www.world-nuclear.org/information-library/nuclear-fuel-cycle/nuclear-power-reactors/appendices/rbmk-reactors.aspx.
14. Mikhail V. Malko, "The Chernobyl Reactor: Design Features and Reasons for Accident," Joint Institute of Power and Nuclear Research, Republic of Belarus, undated. https://www.rri.kyoto-u.ac.jp/NSRG/reports/kr79/kr79pdf/Malko1.pdf.

Figure 79. The aftermath of the Chernobyl explosion. A large amount of radioactive material was thrown up into the atmosphere where it was dispersed over much of Europe and into Russia. The 200-ton roof of the reactor was blown off and the highly radioactive contents of the reactor's core were released. This has never happened in any other nuclear accident.

are still 10 RBMK reactors operating today, all of which are in Russia. Considerable modifications were made to all of them after the accident and they are scheduled to continue operating for some years. The final one that was commissioned in 1990 is scheduled to operate until 2050, and that will be the end of them.[15] It is extremely unlikely that an accident of this magnitude will ever happen again.

For some perspective, 1.35 million people die in roadway accidents every year.[16] In comparison, there have been no more than 60 nuclear-power-related fatalities from the more than 440 nuclear powerplants worldwide; and all of these fatalities were from Chernobyl and the freak accident that occurred because of their poorly designed reactor. That being said, we are more than 1,000,000 times more likely to die in a roadway accident than from a nuclear-power-related accident. But even given these odds, I'm not giving up my car or my bicycle any time soon.

---

15. "RBMK Reactors – Appendix to Nuclear Power Reactors," World Nuclear Association, July 2019. https://www.world-nuclear.org/information-library/nuclear-fuel-cycle/nuclear-power-reactors/appendices/rbmk-reactors.aspx.
16. "Road Traffic Injuries and Deaths – A Global Problem," Centers for Disease Control and Prevention, December 19, 2019. https://www.cdc.gov/injury/features/global-road-safety/index.html.

Fukushima

On March 11, 2011 a rare double earthquake with a magnitude of 9.0 occurred 150 kilometers (90 miles) off the east coast of the island of Honshu, Japan. The resulting tsunami was up to 15-meters (50-feet) high and inundated 560 square kilometers (218 square miles) of land. Nearly 20,000 people died as a result.[17] There are no reports found asserting anyone was killed by the earthquake itself. Japan is a world leader in designing buildings that can withstand powerful earthquakes for a good reason.

When the earthquake hit, three of the six reactors at the Fukushima Dhi-ichi site at seaside were operating at the time and shut down automatically, as did all other operating reactors in the vicinity. The reactors were not damaged by the earthquake, proving they were properly designed for such a powerful event. Four of the six reactors, however, were situated at far too low an elevation, especially given the knowledge of historical tsunamis in Japan.

Nearly one hour after the earthquake, a 14-meter (46-foot) tsunami came ashore at the site of the six nuclear reactors at Fukushima Dhi-ichi. For the following week or so, this was a case of nearly everything going wrong due to bad design and poor decision making.

Other reactors in the region remained connected to the electrical grid during the incident, but Fukushima lost its connection to the electrical grid early on because of damage caused by the earthquake. When a reactor shuts down in an emergency situation it must have cooling water pumped to the core of the reactor continuously to avoid a nuclear-fuel meltdown. This is not due to the nuclear reaction but rather due to the heat of decay of the fission products in the core that are the result of previous reactions.[18] It takes at least four days of continuous cooling, while the highly radioactive fission products decay and become cooler, to prevent a meltdown of the core.

When grid power was lost early during the incident, the emergency back-up diesel generators kicked in to power the pumps sending cooling water to the core. But then an hour later came the tsunami and the beginning of the saga that destroyed three multi-billion-dollar nuclear reactors. The back-up diesel generators were situated on the seaward

---

17. Shinji Nakahara, et al., "Mortality in the 2011 Tsunami in Japan," PubMed, January 5, 2013. https://www.ncbi.nlm.nih.gov/pmc/articles/PMC3700238/.
18. Fission products are the elements, some of them highly radioactive, that result from the splitting of the uranium atoms, which is the "fuel" in the reactor. Many of the fission products decay quite quickly so that after about four days it is no longer necessary to cool the reactor core to prevent a meltdown event.

side of the nuclear plants. They were not secured to the surface in any way and were simply out in the open on skids, as were their fuel tanks. They were immediately swept away leaving the three crippled reactors with no emergency back-up power for cooling the core. The subsequent meltdowns and hydrogen explosions led to the media frenzy that more-or-less predicted the end of life as we know it. Unfortunately, this caused a huge backlash against nuclear energy in the West, as well as in Japan. Except for Japan, the rest of the far East was not so naive to be easily swayed. If the back-up generators had been in bunkers on higher ground with buried cables to the emergency pumps the accident would not have occurred. And if the reactors had been built on higher ground then the emergency pumps would have been right there with them. In fact, reactor units five and six at Fukushima were built on higher ground after the first four units were built on lower ground. The tsunami did not damage units five and six because of where they were situated.

I will never forget the CNN headline on the TV screen in the midst of the accident that declared: "Nuclear Crisis Deepens as Bodies Wash Ashore." It appears to have been removed from the internet, but it seemed to imply that Fukushima was responsible for nearly 20,000 tsunami-caused deaths, while there was actually not one death caused by radiation from this unfortunate accident. In fact, the evacuation of 150,000 people living in the vicinity of the reactors is officially recognized as the cause of 2,259 deaths, including and as a result of evacuating seven intensive-care wards to gymnasiums which were not equipped for that purpose.[19,20] Evan Douple, Associate Chief of Research with the Radiation Effects Research Foundation in Hiroshima was interviewed by Richard Knox of NPR who asked:

> *What do you think about the idea of studying health effects from the Fukushima Dai-ichi accident?* Evan Douple replied, *I think it would be very unwise. There just isn't any evidence that there are enough exposed people at high-enough doses to expect to see any health effects that are measurable.*[21]

---

19. "Fukushima Daiichi Accident," World Nuclear Association, updated May 2020. https://www.world-nuclear.org/information-library/safety-and-security/safety-of-plants/fukushima-daiichi-accident.aspx.
20. Tsukasa Namekata, "Health Consequences of Mandatory Evacuation Due to Nuclear Accident Caused by the 2011 Tsunami," Washington State Public Health Association, October 13-15, 2013. https://washingtonstatepublichealthjournal.files.wordpress.com/2014/02/wspha_presentationnamekata.pdf.
21. Richard Knox, "Why We May Not Learn Much New About Radiation Risks From Fukushima," NPR, March 24, 2011. https://www.npr.org/sections/health-shots/2011/03/24/134833008/why-we-may-not-learn-much-new-about-radiation-risks-from-fukushima.

The Radiation Effects Research Foundation (initially named the Joint Commission for the Investigation of the Effects of the Atomic Bomb in Japan) was set up jointly by Japan and the United States after World War II to study the health history of the survivors of the atomic bombs dropped on Hiroshima and Nagasaki. For the study they included: 120,321 survivors; 26,580 residents of the two cities who were away at the time of the bombing; 3,289 women who were pregnant at the time of the bombings; and 76,814 people who were born after the bombings, at least one of whose parents were exposed to radiation from the bombs.[22]

One of the reasons this is such an important study is that the researchers know exactly where the bombs were detonated and that they were a single-point source of radiation. They also know where each survivor was located and whether they were out in the open or shielded by wood, concrete, or some other material. As a result, they can quite accurately determine the dose of radiation each subject received and from there deduce the dose that causes various types of sub-lethal health effects. This research supports the hypothesis that below a certain level of exposure to nuclear radiation there is no harmful effect. It also confirms that people receiving a very high sub-lethal dose of radiation are slightly more likely to develop cancers later in their lives.

Nuclear Energy – The Only Technology that can Replace Fossil Fuels

I do not believe that the demonization of fossil fuels and carbon dioxide are in any way justified. But I do believe in the conservation of important limited resources if an alternative, that is economically and technically feasible, can be found. And I do not believe that wind and solar with battery back-up is economically or technically viable to replace a high percentage of fossil fuels on a global scale. Enough about my beliefs, here's what nuclear energy could do for stretching out the lifetime of fossil fuels without crippling the economy.

China, Russia, and India are now the world leaders in building nuclear powerplants to produce electricity. They are reliable, close to zero emissions, and unlike wind energy they keep running even when there is no wind, and unlike solar energy they keep running even when it's cloudy and dark. Today the United States has 95 operating reactors, two under construction, and three are being planned. France has 56

---

22. Radiation Effects Research Foundation. https://www.rerf.or.jp/en/

operating, one under construction, and zero being planned. China has 48 operating, 12 under construction, and 44 being planned. Russia has 38 operating, four under construction, and 24 being planned. India has 22 operating, seven under construction, and 14 being planned.[23]

As an example of how effective nuclear energy is at reducing the need for fossil fuels, France produces 71 percent of its electricity with nuclear energy and emits five tons of carbon dioxide per person every year, whereas Germany produces only 11 percent of its electricity from nuclear energy and emits 9.1 tons of carbon dioxide per person every year. If Germany shuts down its seven remaining nuclear plants by 2022 as planned, its CO_2 emissions will inevitably rise.

Eleven countries produce more than one-third of their electricity with nuclear energy. Those countries are Belgium, Bulgaria, the Czech Republic, Finland, France, Hungary, Slovakia, Slovenia, Sweden, and Ukraine. Every country in the world could achieve this level and higher if it was set as a priority or a goal. Nuclear plants create a guaranteed income stream from a large base of electricity consumers, so it is mainly a case of political will and investor confidence.

Figure 80. Two of the four reactors at South Korea's Shin Kori nuclear park in the southeast of the country. Each one is capable of producing 1,400 megawatts of electricity, so these two reactors alone can power nearly two million homes. South Korea has 24 nuclear reactors producing 30 percent of their electricity with four more under construction. The South Korean nuclear industry prides itself in making their reactors look nice to the eye. Inside many of the walls and floors are surfaced with marble, something not common in other countries where bare concrete is considered sufficient.

---

23. "World Nuclear Power Reactors & Uranium Requirements," World Nuclear Association, September 2020. https://www.world-nuclear.org/information-library/facts-and-figures/world-nuclear-power-reactors-and-uranium-requireme.aspx.

The key to reducing fossil fuel consumption, if that is the goal, is to electrify our economies with nuclear energy. Some things are easy to do and others very difficult, verging on impossible. But my proposed program is not based on fear of CO_2 and the so-called, and impossible, "net-zero" goal of the catastrophists in the environmental movement. My proposed program is much more practical with the parallel goals of diversifying energy technologies and conserving limited resources, so they last longer. Nuclear can do this better than any other known energy technology.

The most obvious starting point is to operate all buildings with nuclear electricity. Nuclear could provide energy for heating, cooling, hot water, lighting, all appliances, and other electrical devices. If I had a say, I would make an exemption for cooking with gas – it's a very small part of fossil fuel energy use.

In the United States, buildings consume 40 percent of the total energy production, more than transportation or industry.[24] This shift would be easy. Air conditioning is already 100 percent electric, as are appliances, tools, etc. Heating with electricity can be done with a heat pump, with baseboard heaters on individual thermostats in each room, or by in-floor (hydronic) heating with individual thermostats as well.

Industry already uses a lot of electricity, but the majority of that electricity is produced using fossil fuels in most countries today. Nuclear energy could also provide a large percentage of the electricity for mining, manufacturing, and energy-intensive technologies such as electric-arc furnaces for recycling steel. In fact, nearly every requirement for energy that is stationary, as opposed to transport and heavy mobile machinery, could be provided by nuclear energy.

Transportation on land and in the air is by far the biggest challenge when it comes to finding alternatives to fossil fuels. One exception to this is that all trains could be electrified as a majority of them have been in many European countries and in Japan.[25,26] It is still questionable whether electric passenger cars, which could be recharged with nuclear electricity, will displace the internal combustion engine in the majority of vehicles. Heavy transport and other heavy equipment used for excavating, road building, etc. is even more problematic. Aircraft

---

24. "Energy Use in Buildings," Alliance to Save Energy, 2018. https://www.ase.org/initiatives/buildings https://www1.eere.energy.gov/buildings/publications/pdfs/corporate/bt_stateindustry.pdf.
25. "Percentage of the railway lines in use in Europe in 2017 which were electrified, by country," Statista, May 5, 2020. https://www.statista.com/statistics/451522/share-of-the-rail-network-which-is-electrified-in-europe/.
26. "Rail Transport in Japan," Wikipedia, September 5, 2020. https://en.wikipedia.org/wiki/Rail_transport_in_Japan.

Figure 81. The Russian nuclear-powered icebreaker Yamal, one of five on active duty - primarily to keep the shipping route from Murmansk to Vladivostok open during the winter months. If icebreakers and submarines carrying thermonuclear weapons can be nuclear powered, then all deep-sea ships can be nuclear-powered.

are the greatest challenge of all. It is unlikely they will be converted to liquid hydrogen and liquid oxygen – as are used for most spacecraft – anytime soon.

While land and air transport are a real challenge, the sea is quite hospitable to nuclear-powered ships.[27] The fleet of five Russian icebreakers in the Arctic is nuclear powered and there are three more currently under construction (see Fig. 81). Six countries have nuclear navies, including aircraft carriers, battleships, cruisers, and submarines.[28] This indicates that it is feasible to operate all deep-sea shipping with nuclear energy.

Unlike the climate alarmists who want to displace fossil fuels almost immediately, because they think this will save the world from a climate disaster, it makes more sense to take a practical approach that would phase in nuclear energy where it is most cost-effective; thus reducing our heavy reliance on fossil fuels incrementally. This is the ambition in Russia, China, and India today.

One of the most irrational aspects of the climate alarmism movement is that the vast majority of the people in the movement are

---

27. "Nuclear Marine Propulsion," Wikipedia, September 2, 2020. https://en.wikipedia.org/wiki/Nuclear_marine_propulsion#Icebreakers.
28. "Nuclear Powered Ships," World Nuclear Association, July 2020. https://www.world-nuclear.org/information-library/non-power-nuclear-applications/transport/nuclear-powered-ships.aspx.

adamantly opposed to nuclear energy. This is the case, even though it could obviously displace far more fossil fuel that any other technology. There are notable exceptions to this position such as Jim Hansen of NASA who could be called the father of climate alarmism, yet he supports nuclear energy.[29] Michael Schellenberger, who was a climate alarmist but supported nuclear energy, has now decided that alarmism is not the right way to make environmental progress, not even about climate change.[30] There is hope, but there is not yet a universal recognition of the fact that the climate is not in despair, and that fossil fuels are a big factor for our economic well-being and longevity, and moreover that carbon dioxide is the foundation of all life. Let's hope real science and reason prevail against the current epidemic of ideology, misinformation, and fear.

---

29. Thom Mitchell, "'Father of Climate Change' James Hansen Urges Support For Nuclear Energy At #COP21 Climate Talks," NewMatilda.com, December 4, 2015. https://newmatilda.com/2015/12/04/father-of-climate-change-james-hansen-urges-support-for-nuclear-energy-at-cop21-climate-talks/.
30. Michael Schellenberger, *Apocalypse Never: Why Environmental Alarmism Hurts Us All*, Harper: Illustrated Edition, June 30 ,2020. https://www.amazon.com/Apocalypse-Never-Environmental-Alarmism-Hurts/dp/0063001691.

CHAPTER 9

Forest Fires: Of Course They are Caused by Climate Change (Not Trees?)

There are three primary causes of wildfires, including forest fires, brush fires, and grass fires. Those causes are: lightning, fires caused accidently by humans – such as by campfires, powerlines, and cigarettes – and fires caused on purpose by arsonists or by forest managers using controlled (prescribed) burning techniques as a forest management tool.

Little can be gained by arguing about whether forest fires in general are good or bad. Firstly, forest fires come in a great variety of sizes and intensities. Some fires burn a small area and kill only the shrubs and groundcover, leaving the trees alive. Other fires kill virtually everything over vast areas including the seeds and soil, leaving the site of the fire sterile and subject to erosion. Secondly, while forest fires are often beneficial, as a way of temporarily increasing forage for wild grazing animals, they are just as often harmful for soil, trees, fish, birds, and humans. We tend to think worse of the wildfires that have the largest insurance claims, when there is loss of human life and property. This may seem reasonable from a human perspective, but it has little to do with the health of forest ecosystems. But there is a common solution to alleviate both of these circumstances: wise management of forests based on an understanding of forest ecology and implementing specific techniques to reduce the risk and severity of wildfires.

Lately it has become the fashion to blame climate change for forest fires. In other words, to blame rising levels of carbon dioxide in the atmosphere for what is causing forest fires. Here is the historical record of land burned by wildfires in the United States since 1926 (see Fig. 82). It would appear that the impact of climate change has moderated somewhat since the 1930s, long before CO_2 emissions rose to the magnitude they are today. The claim that climate change is causing the forest fires, mainly in the West, is a dereliction of duty on the part of politicians and "green" activists. If they really were advocates for a green earth, they would take some courses in forest ecology and study the long history of droughts in the West going back many centuries.

In 1924 the US Congress passed the Clarke-McNary Act, an agreement among forest landowners, the western States, and the Federal

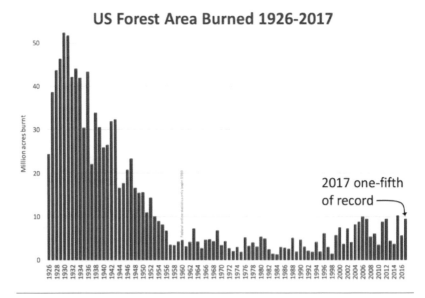

Figure 82. Here is a graph showing the amount of forest acres burned in the United States from 1926 to 2016 (the reference below has data up to 2019). Notice the sharp decline in the 1940s when fire suppression became the precedent for forest management and then the gradual rise from the mid-1990s when the environmental movement began to have more influence, promoting a "hands-off" management approach. This graph illustrates important trends in forest management during the past 95 years.*

* National Interagency Fire Center, "Total Wildland Fires and Acres (1926-2019)," 2020. https://www.nifc.gov/fireInfo/fireInfo_stats_totalFires.html.

government to cooperate in controlling fires.[1] A monument at Snoqualmie Falls in Washington State commemorates this historical meeting where the agreement was reached. Since that time, fighting fires has become a sophisticated enterprise employing satellite surveillance, helicopters, fire-retardant chemicals, and water bombers, in addition to the traditional fire-spotters in mountaintop watchtowers. Each year thousands of fires are reported with most of them being controlled before they spread very far. Some of them get away however, and when they do, they cause a lot of damage before they are contained.

As a result of the Clarke-McNary Act, aggressive fire suppression became the primary policy of the US Forest Service in cooperation with state governments and landowners. Smokey the Bear was created in 1944 with the slogan "only you can prevent forest fires." This did have the effect of drastically reducing the area of land burned annually. But this also ensured the buildup of dry fuel load in the form of dead grass, brush, tree limbs, and dead trees. There are two main types of fires in a forest. Ground fires tend to leave the trees relatively intact. But there are also crown fires where, due to a heavy fuel load on the ground, the fires spread into the tops of the trees. If the trees are growing close together the wind can cause an inferno and the fire can travel faster than any human or wildlife can run.

It is both instructive and important to note that most of the biggest and most damaging wildfires occur on public lands such as National Forests and National Parks.[2] Unlike public land, private forest lands are managed to prevent catastrophic fires by proactive thinning and fuel load reduction, and by building road access to get firefighters in on the ground quickly. For example, the US Southeast produces more lumber than anywhere else in the country, most of its forest lands are privately owned and managed. Those lands are mostly pine forests, as is the case in much of the West. But there are far fewer forest fires, particularly large fires, in the Southeast than in the West.

From the mid-1950s to the mid-1990s the policy of fire suppression kept the annual area burned relatively low. However, during this time the fuel load continued to accumulate on public lands, and by this time the green movement, which is largely urban based, gained support for their philosophy that forest fires are natural and should be allowed to

---

1. Lewis F. Southard, "The History of Cooperative Forest Fire Control," *Forest History Today*, Spring/Fall 2011. https://foresthistory.org/wp-content/uploads/2017/01/Cooperative_Fire_Control.pdf.
2. Melvin Thornton, "Private forest land is not the problem," *Mail Tribune*, December 23, 2018. https://mailtribune.com/opinion/guest-opinions/private-forest-land-is-not-the-problem.

burn unless they threaten lives and buildings.[3] This brings us to the urban-forest interface.

There are a number of other reasons why forest fires are more prevalent in the West, one of which is that the West is more drought-prone than the East. Another is that the trees in the West are predominantly coniferous trees while in much of the East, especially the Northeast, the trees are predominantly broadleaf trees of the Carolinian hardwood forest. Coniferous trees contain a lot of pitch which makes them burn more like a candle whereas broadleaf trees such as maples and oaks do not have a lot of pitch in their leaves or their bark and are more resistant to fire. Residential areas should never be placed into a landscape of coniferous trees, period. This was the case in Paradise, California where in 2018 the Camp Fire destroyed 95 percent of the community and the official death toll was 86.[4] The fire was caused by a Pacific Gas and Electric (PG&E) powerline and the residences were essentially built in the forest where a combination of ground fires and crown fires razed whole sections of housing to the ground (see Fig. 83). In addition it was confirmed that PG&E had not kept their powerlines safe from contact with surrounding trees.

When the Australian aboriginal people arrived there about 60,000 years ago, they found a land rich in wildlife, forests, and grasslands. It didn't take them 60,000 years to learn how to manage the land to help prevent catastrophic wildfires. They used "fire-stick" farming, to set fire to grasslands and forested lands before the summer, to burn off the dead grass and wood to reduce the fuel load.[5] Then if a fire did start in the hot and dry season it would be much less severe. When Native Americans arrived in the New World over the land bridge about 15,000 years ago, they also had a good amount of time to figure out similar practices. The trick is to light the fires before it gets too hot and dry and when high winds are not prevailing. In a landscape that is lightly populated it is usually possible to do this without burning down a settlement. Today this can only be practiced in regions that are somewhat remote from human populations.

Even the most carefully planned program of controlled burning can go wrong, and if the wind comes up or changes direction unexpectedly,

---

3. Edward Ring, "Environmentalists Destroyed California's Forests," California Policy Center, September 10, 2020. https://californiapolicycenter.org/environmentalists-destroyed-californias-forests/.
4. Associated Press, "Death toll in Camp Fire probably includes 50 more people, report says," Los Angeles Times, February 11, 2020. https://www.latimes.com/california/story/2020-02-11/death-toll-in-camp-fire-probably-includes-50-more-people-report-says.
5. D. M. J. S. Bowman, "The impact of Aboriginal landscape burning on the Australian biota," New Phytology 140 pp385-410, 1998. https://nph.onlinelibrary.wiley.com/doi/pdf/10.1111/j.1469-8137.1998.00289.x.

Figure 83. Part of Paradise, California after the Camp Fire in November 2018. This is the sad result of not understanding forest ecology. Closely packed coniferous trees should never be this close to buildings. In a landscape like this, most of the coniferous trees should have been removed and used to make lumber and paper. Then, where trees were desired for landscaping, broadleaved trees should have been planted, well-spaced with lots of open parkland, and with regular removal of dead wood.

this can result in damage to property and loss of life. The people who set these fires with good intentions were not very popular with the surrounding townsfolk in the aftermath. In regions that are close to villages and towns, it makes more sense to use mechanical means to thin the forests and remove dead wood.

These fire prevention programs cost money. If there is no income derived from the forest, it becomes difficult to find the budget to properly manage the forest. It seems a logical objective is to manage some of the public forest for timber production, thus earning some income, which can then be used to manage fuel loads and produce a sustainable harvest of timber. New trees will either grow back on their own or can be replaced by trees planted or regenerated in the managed plantations. But the National Forests in the United States, which were established with the intention of "multiple use" including timber harvesting, have seen a drastic decline in timber harvesting and thus, have seen a drastic decline in income that could be used for forest management (see Fig. 84, page 166).

There are 421 sites totaling 86.4-million acres (35-million hectares) managed by the US National Parks Service. This amounts to 3.4 percent of US land. This land cannot be used for commercial timber production, but perhaps some of the revenue from other public lands, including the National Forests and Bureau of Land Management land could be used to manage National Parks in a manner that would reduce wildfire risk.

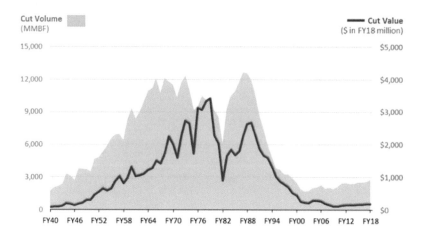

Figure 84. Timber harvest cut and value from National Forest lands from 1940 to 2018. Undue concern for the spotted owl and anti-forestry campaigns reduced the annual cut to less than two percent of its peak in the mid-1970s. A program of thinning and some actual commercial timber harvest could provide the budget to reduce the fuel load and thereby reduce the risk of large destructive wildfires. The National Forests were not intended to be National Parks. They were designated for multiple functions, including forestry.

Unlike the Australian aborigines and the Native Americans, the people who came to both these regions from Europe, Asia, and Africa beginning in 1492 have only lived in these environments for a little more than 500 years. Surely, we must learn from the wisdom of the first people and incorporate these practices into our modern systems of land management. There are too many people living in or adjacent to forested lands that present a fire risk to ignore the true cause – high fuel load and lack of forest management.[6,7]

An interesting comparison is that there are not a lot of massive wildfires in European landscapes even though the European Union countries have 42 percent forest cover[8] compared to the United States forest cover of 34 percent.[9] It is imperative that the US Forest Service, and the other rural wildland managers within the United

---

6. Jim Steele, "Reducing the Devastation of California Wildfires," CO_2 Coalition, August 27, 2020. http://co2coalition.org/publications/reducing-the-devastation-of-california-wildfires/.
7. Jim Steele and Genesis Torres, "Five Ways to Reduce Wildfire Risk in California," CO_2 Coalition, January 28, 2020. http://co2coalition.org/publications/5-ways-to-reduce-wildfire-risk-in-california/.
8. "Over 40 percent of the EU covered with forests," Eurostat, 2015. https://ec.europa.eu/eurostat/web/products-eurostat-news/-/EDN-20180321-1.
9. https://tradingeconomics.com/united-states/forest-area-percent-of-land-area-wb-data.html.

States and Australia, look to other jurisdictions where there are forest and wildlands plans and policies that substantially reduce the number of dangerous wildfires. It is not rocket science, but it does require knowledge of how forests and landscapes actually work rather that an illusory fantasy based on interpretations of nature that are not based in fact. And no, forest fires are not caused by climate change.

The Use of Wood to Produce Electricity

There has been much criticism of late regarding the use of wood (biomass) to generate electricity. Of course, before fossil fuels were harnessed, wood was by far the most important source for both energy and building material since fire was utilized for cooking and heating, and since timber was first hewn thousands of years ago. That was a long time ago, near the beginning of human civilization. Wood is still the most important source of renewable energy, along with hydroelectric power; and wood is still, by far, the most important renewable building material for construction, furniture, etc. In other words, wood is the most important renewable material substance in the world. Its production is powered by the Sun.

Agriculture began near the beginning of the Holocene 10,000 years ago. Prior to that, humans were in relatively small nomadic groups hunting and gathering from the wilderness during seasonal migrations. Agriculture allowed permanent settlements to develop and towns and cities emerged. These settlements were dependent on agriculture to feed the populace. But forests were so abundant and the human population so small, that there was no incentive to farm trees until the industrial revolution in the early 1800s. By this time the forests of central Europe were shrinking due to reliance on wood for cooking, heating, glassworks, smelting iron and copper, and fueling steam engines. At their low point, European forests were reduced to less than 10 percent of Europe's landscape. This resulted in the development of silviculture, commonly known as forestry, which introduced the farming of trees, to increase the wood supply. Today, 42 percent of Europe is forested due to the success of forestry. Close to 100 percent of these forests are occupied with native tree species, as is also the case in North America.

Over the years since Europeans pioneered silviculture it has spread around the world. Every industrialized country is now practicing sustainable forestry, meaning that new trees are growing back as fast or faster than the mature trees that are being harvested. For example, the forest areas of both Canada and the United States are higher today than they were in 1900.[10] This is in part due to the care taken of forests, the demand for wood, and the intensive agricultural practices that require less land to produce more food. This is an important point. The more food we can produce from a given area of farmland, the more land there is that can remain forested, for future timber production or for preservation. There is only so much land that is suitable for farming and forestry. Increased atmospheric carbon dioxide will enhance both the farming and forestry industries as a result of the CO_2-fertilization effect.

Let's revisit the current opposition to burning wood in state-of-the-art powerplants as a source of energy to produce electricity; it is a simple subject. So long as the wood used to produce electricity is being used at a sustainable rate, and is comparable to the growth rate of forests, it doesn't really matter what the wood is used for. What does matter is that the wood used for construction, furniture, heating, or producing electricity is not being used at a faster rate than it is growing. This is the very definition of sustainability. Of course, as with any combustion process, state-of-the-art pollution-control technology must be employed.

The wood that is being used to produce electricity is wood that is not suitable for the production of construction lumber. Construction lumber will always be more valuable than the wood chips and sawdust used to make wood pellets for energy production. So, the wood used for energy will always be a by-product of the trees cut for solid-wood production, or in many tropical countries, for paper production.

It is still a surprise to some people that the more wood used, the more trees will be planted and thus, the overall forested area of a region or country will be increased. Take the country of Haiti, for example; in Haiti they build nearly all their structures out of concrete. Therefore, they don't need wood and they have not planted new forests to replace the ones they cut down for firewood many years ago. The country has been practically denuded of forests. If they established forests, they could then build their houses with wood and the forests would be maintained into the future to satisfy the ongoing demand for wood. Using more wood equals growing more trees.

I would not be surprised if the wind and solar energy lobbyists are behind these recent efforts to denigrate the use of wood in producing electricity. Biomass, including wood and other combustible materials, accounts for a much larger percentage of energy produced than wind

---

10. Douglas W. MacCleery, "American Forests – A History of Resilience and Recovery," Forest History Society, 2011.

Figure 85. A truck delivers wood chips to a state-of-the-art biomass powerplant in Virginia. Globally, biomass energy produces much more electricity than wind and solar combined; and unlike wind and solar energy, biomass can be counted on to deliver electricity when it is needed.

and solar combined. The state of Virginia produces four percent of its electricity from biomass, the majority of which is wood by-product from sawmills and forestry operations (see Fig. 85).[11] The state of Vermont produces five percent of its electricity from wood waste.[12] If the wind and solar lobbyists can turn people against biomass, their niche for renewables, granted unreliable and costly technologies, would be the alternative. Biomass energy production also has the distinct advantage of having the ability of operating continuously and therefore does not need an entire back-up system to compensate for intermittency as do wind and solar energy.

Wind and solar energy production continue to disappoint. Neither solar nor wind generation are available even 50 percent of the time. Solar energy averages between 10 to 20 percent capacity compared to what it would produce if operating at full power 24 hours per day. There is no power at night when you want light and there's little production when it's cloudy during the daytime or during the early or late periods of the day even when it is sunny. Wind averages between 20 to 30 percent capacity.[13] When the wind isn't blowing, some other source

---

11. "Biomass in Virginia." http://www.virginiaplaces.org/energy/biomass.html.
12. "Vermont – Select economic and energy data," Institute for Energy Research. https://www.instituteforenergyresearch.org/states/vermont/.
13. Nicholas Boccard, "Capacity Factor of Wind Power: Realized Values vs. Estimates," SSRN, January 19,

of electricity must be available to carry the load. This is what makes solar and wind so much more expensive than other more reliable and efficient technologies. This is also why California and Australia are experiencing blackouts, and why so many back-up diesel generators are being installed in hospitals and other buildings that cannot cope with long power outages.

Let's look at two recent examples of articles with a strong "anti-biomass" message. The first is from a website called Eco-Business, where they published an article titled: "Are forests the new coal? Global alarm sounds with biomass burning."[14] The article begins:

> *The forest biomass industry is sprawling and spreading globally – rapidly growing in size, scale, revenue, and political influence – even as forest ecologists and climatologists warn that the industry is putting the planet's temperate and tropical forests at risk, and aggressively lobbying governments against using wood pellets as a 'renewable energy' alternative to burning coal.*

So, energy produced by burning renewable wood is not renewable energy? No "forest ecologist" would ever make such a statement, and probably why no one is mentioned by name in the article. Although, "climatologist" is probably shorthand for "climate alarmist." The article goes on to quote an environmental academic:

> *Michael Norton, environmental director of the Science Advisory Council of the European Academies, said in a December 2019 statement issued to European Union countries, 'the reason is simple: when the forest is harvested and used for bioenergy, all the carbon in the biomass enters the atmosphere very quickly, but it will not be reabsorbed by new trees for decades. This is not compatible with the need to tackle the climate crisis urgently,' said Norton.*

This is nonsense. Forests are managed in what's called a "rotation." A rotation is the time between planting new tree seedlings in a harvested area and when those trees are eventually fully harvested. The rotation time varies due to species, climate, and intended use of the wood. Consider a forest landscape managed on a 50-year rotation. This means every year 1/50th of the total area can be harvested while

---

2016. https://papers.ssrn.com/sol3/papers.cfm?abstract_id=1285435.
14. Justin Catanoso, "Are forests the new coal? Global alarm sounds with biomass burning," Eco-Business, September 1, 2020. https://www.eco-business.com/news/are-forests-the-new-coal-global-alarm-sounds-with-biomass-burning.

49/50ths of the forest are continuing to grow, all the while absorbing carbon dioxide. This is why the Intergovernmental Panel on Climate Change (IPCC) considers the forest industry "carbon neutral," because the growing trees are absorbing CO_2 as fast as the trees are being harvested. The fuel that is used for operating machinery in the forest and transporting wood is counted separately because it does not replace itself like trees do.

The article continues and characterizes the IPCC's position as a "loophole":

> Today's forest biomass industry is both refuting and shrugging off its environmental critics and appears to be on a roll. That's largely thanks to the so-called United Nations' carbon accounting loophole that designates the burning of forests to generate electricity as carbon-neutral, despite recent hard science that shows otherwise.

This reference to "hard science" is from an article in *Science*, the journal of the American Association for the Advancement of Science, a publication that long ago dropped any pretext of scientific rigor. They frequently publish articles that are as sensationalist as they are just plain wrong. This is one such article.[15]

Another article worth pointing out is one published on the website NoTricksZone that contains the headline: "Environmental Disaster: Northern Europe Deforestation Up 49 percent Due To Effort To Meet CO_2 Targets."

Here we have a classic case of confusing deforestation with forestry. Deforestation is when forested land is converted to some other use permanently. Examples of deforestation would be removing a forest and using the land for agriculture, housing, or industry. Felling trees is a somewhat integral part of forestry and should not be confused with deforestation. The article states:

> Germany's flagship ARD public broadcasting presented a report earlier today about how 'CO_2 neutral' wood burning is leading to widespread deforestation across northern Europe…The ARD's 'Das Erste' reports how satellite images show deforestation has risen 49 percent since 2016 in Sweden, Finland, and the Baltic countries. The reason: 'Because of the CO_2 targets. That sounds

---

[15]. Timothy D. Searchinger, "Fixing a Critical Climate Accounting Error," *Science*, October 23, 2009. https://science.sciencemag.org/content/326/5952/527.summary.

totally crazy but precisely because of the trend to renewable energies is in part responsible for deforestation in Estonia,' says the Das Erste moderator.[16]

The 49 percent rise in "deforestation" has supposedly been detected by satellite but there is no sign of it when zooming in with Google Earth Pro. All the Nordic and Baltic countries continue to appear nearly 100 percent green, with farms being a lighter green and mainly spruce and pine forests being darker green. It seems the 49 percent figure in the article is entirely fictitious.

As with the benefits of using unrecyclable, combustible municipal waste for fuel, there is also great benefit to using wood – the wood not suitable for lumber – as fuel to make electricity or heat. This can include wood products such as bark, sawdust, chips, as well as forest-floor residue.

---

16. P. Gossilin, "Environmental Disaster: Northern Europe Deforestation Up 49 percent Due To Effort To Meet 'CO$_2$ Targets,'" NoTricksZone, September 6, 2020. https://notrickszone.com/2020/09/06/environmental-disaster-northern-europe-deforestation-up-49-due-to-effort-to-meet-co2-targets/.

CHAPTER 10

Ocean Acidification – A Complete Fabrication

Although there are a few references earlier in the literature, it wasn't until 2003 that we saw the explosion of journal articles, media reports, and glossy publications from environmental groups on the subject of "ocean acidification" begin to appear. This just happened to coincide with a paper published in the journal *Nature*, which reported this quote below about human emissions of carbon dioxide:

> *May result in larger pH changes [in the oceans] over the next several centuries than any inferred from the geological record of the past 300 million years.*[1]

I apologize to the reader for going over the carbon dioxide history once again, but this chapter would have a hard time being complete if we were not to mention a little bit more about CO_2. As with so many of these fake catastrophes, this one is also blamed on the indiscernible carbon dioxide, which is just as invisible in the sea as it is in the air.

---

1. Ken Caldeira and Michael E. Wickett, "Anthropogenic carbon and ocean pH," *Nature* 425 (6956) (2003): pp365-365. September 2003. https://www.nature.com/articles/425365a.

The ocean acidification hypothesis proposes that increases in atmospheric levels of CO_2 will inevitably result in the oceans becoming more acidic as they absorb more carbon dioxide, some of which reacts in the sea to become carbonic acid. In turn, it is proposed this would result in a lowering of oceanic pH which would cause a serious, even "catastrophic" effect on shellfish, plankton, and corals. It would affect all calcifying marine species, that is any species that builds protective shells of calcium carbonate from calcium and carbon dioxide dissolved in the seawater. The projected lowering of the ocean's pH would then make it difficult or even impossible for these species to construct their shells, and thus, some researchers have claimed they will become extinct.

The term ocean acidification is, in itself, very misleading. The scale of pH runs from zero to fourteen, where seven is neutral, below seven is acidic, and above seven is basic, or alkaline. The pH of the world's oceans varies from 7.5 to 8.3; notice that this is well into the alkaline scale. Nowhere is it acidic. But proponents of this hypothesis claim the oceans will become "more acidic" when there is no evidence of them being acidic in the first place. It is scientifically incorrect to use the language in this manner. It is also well known that the terms "acid" and "acidic" have strong negative connotations for most people, whereas "basic" and "alkaline" do not.

It would certainly be of dire consequence if human emissions of carbon dioxide were to kill all the clams, oysters, snails, crabs, shrimp, lobsters, coral reefs, and the many other calcifying species in the oceans along with all the species that depend on them for food. This chapter will examine this hypothesis in detail and test its assumptions against real-world observations and actual scientific knowledge.

For those campaigning on the issue of climate change, the specter of ocean acidification neatly solves the problem created by the failure of the global temperature to rise at the rate predicted, due to of course, rising CO_2 in the atmosphere. This was especially true during the "pause" in warming between 1995 and 2014 (see Fig. 86).[2] The hypothesis of ocean acidification does not require any warming, any change in climate, or any increase in extreme weather events to occur. It is solely based on the contention that higher levels of carbon dioxide in the atmosphere will result in a lowering of ocean pH, which in turn will cause the extinction of shellfish and all the other calcifying marine species.

---

2. Ross R. McKitrick, "HAC-Robust Measurement of the Duration of a Trendless Subsample in a Global Climate Time Series," *Open Journal of Statistics* 4, pp527-535, 2014. https://www.scirp.org/journal/paperinformation.aspx?paperid=49307.

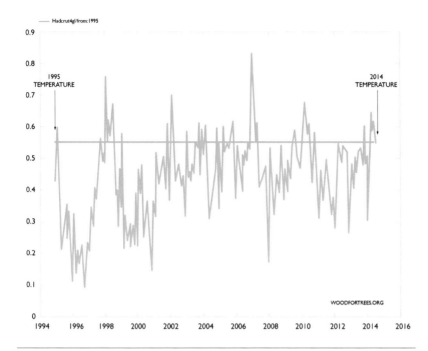

Figure 86. The 19-year pause in temperature rise, coincident with the highest human-caused CO_2 emissions in history at that time. It was also coincident with the invention of the ocean acidification "catastrophe." The data is provided by HadCRUT4 which is developed by the Climatic Research Unit (University of East Anglia) in conjunction with the Hadley Centre (UK Met Office).

Ocean acidification has been dubbed "global warming's evil twin," thus injecting a degree of morality into the discussion.[3] It is disconcerting that scientists, whom one might expect would be more moderate in their tone, employ such alarming language when the future is not yet known. An example from the journal *Trends in Ecology and Evolution*:

> *The anthropogenic rise in atmospheric CO_2 is driving fundamental and unprecedented changes in the chemistry of the oceans. ...We argue that ocean conditions are already more extreme than those experienced by marine organisms and ecosystems for millions of years, emphasizing the urgent need to adopt policies that drastically reduce CO_2 emissions.*

---

3. Carles Pelejero, et al., "Paleo-perspectives on ocean acidification," *Trends in Ecology and Evolution* 25, no. 6, pp332-344, 2010. https://www.cell.com/trends/ecology-evolution/fulltext/S0169-5347(10)00044-3.

Yet, the same paper states: "Understanding the implications of these changes in seawater chemistry for marine organisms and ecosystems is in its infancy."[4]

By 2009, the Natural Resources Defense Council (NRDC), an environmental advocacy group, was predicting that "by mid-century... coral reefs will cease to grow and even begin to dissolve" and ocean acidification will "impact commercial fisheries world-wide, threatening a food source for hundreds of millions of people as well as a multi-billion-dollar industry."[5]

Therefore, not only are calcifying species threatened, but also the entire web of life in the seas. The NRDC document also contended:

> *Acidification may already be impacting marine life around the word. For example, Pacific oysters have not successfully reproduced in the wild since 2004.*

This is clearly not true, as Pacific oyster production has been increasing steadily from 156 thousand tons in 1950 to more than 4.4 million tons in 2016.[6] This assertion is false as much of the Pacific oyster production is based on the collection of wild seed (spat). As stated in the report referenced above:

> *Much of the global supply of spat is obtained from wild seed capture, using a wide variety of settlement materials (cultch) hanging in suspension from long-lines and rafts. However, other commercial units operate hatcheries* (to obtain their seed).

It should also be mentioned that the seed produced in hatcheries is spawned by oysters growing in seawater pumped from the ocean.

The proponents of ocean acidification think that the oceans are absorbing 30 to 50 percent of human CO_2 emissions.[7] A 30-year study near Bermuda contains evidence that the oceans are absorbing carbon dioxide from the atmosphere.[8] Another study based on direct observation in

---

4. Ibid.
5. "Ocean Acidification: The Other CO_2 Problem," Natural Resources Defense Council, August 2009. https://www.nrdc.org/oceans/acidification/files/NRDC-OceanAcidFSWeb.pdf.
6. "The State of World Fisheries and Aquaculture 2018," UN Food and Agriculture Organization, 2018. http://www.fao.org/3/i9540en/i9540en.pdf.
7. John Pickrell, "Oceans Found to Absorb Half of All Man-Made Carbon Dioxide," National Geographic News, July 15, 2004. http://news.nationalgeographic.com/news/2004/07/0715_040715_oceancarbon.html.
8. N. R. Bates, et al., "Detecting anthropogenic carbon dioxide uptake and ocean acidification in the North Atlantic Ocean," *Biogeosciences* 9, no. 7: pp2509-2522, 2012. http://www.bios.edu/uploads/Batesetal2012bg-9-2509-2012.pdf.

the field indicated that increasingly the CO_2 that was not showing up in the atmosphere was being absorbed by biomass on the land.

> *Increasing trends in carbon uptake over the period 1995–2008 are nearly unanimously placed in the terrestrial biosphere (assuming fossil fuel trends are correct), with a small ocean increase only present in a few inversions. The atmospheric CO_2 network is probably not yet dense enough to confirm or invalidate the increased global ocean carbon uptake, estimated from ocean measurements or ocean models.*[9]

There is direct evidence that trees and plants are taking up a large percentage of human CO_2 emissions as their levels increase.[10,11,12] The "CO_2 fertilization effect" is well documented. The fact that greenhouse growers around the world purposely increase the level of carbon dioxide in their greenhouses in order to increase the yield of their crops by up to 50 percent supports this very fact. There is simply no question that terrestrial plants benefit greatly from the increased levels of carbon dioxide in the air. Whereas CO_2 is now at about 415 ppm, the maximum growth rate of most plants occurs at 1,000 to 2,000 ppm, and in some species, it is even higher. The precise division between carbon dioxide absorbed by the oceans and the carbon dioxide contributing to increased terrestrial biomass is very difficult to determine.

This chapter will consider five factors that bring into question the assertion that ocean acidification is a crisis that threatens all or most calcifying species, as well many other species, with extinction. It is important to recognize the role that apocalyptic language plays in this discussion. It could truthfully be said that, "Human emissions of CO_2 may result in a slight reduction of ocean pH that is well below historical levels during which calcifying species have survived and even flourished." Or it could also be stated that, "Global warming and ocean acidification will result in the death of all coral reefs by 2050, resulting in the extinction of thousands of species as the marine environment is

---

9. P. Peylin, et al., "Global atmospheric carbon budget: results from an ensemble of atmospheric CO_2 inversions," *Biogeosciences* 10, no. 10 pp6699-6720, 2013. https://bg.copernicus.org/articles/10/6699/2013/bg-10-6699-2013.pdf.
10. R. J. Donohue, et al., "Impact of CO_2 fertilization on maximum foliage cover across the globe's warm, arid environments," *Geophysical Research Letters* 40, no. 12, pp3031-3035, 2013. https://agupubs.onlinelibrary.wiley.com/doi/full/10.1002/grl.50563.
11. Smithsonian Institution, "*Forests are growing faster, ecologists discover; Climate change appears to be driving accelerated growth,*" ScienceDaily, February 2, 2010. http://www.sciencedaily.com/releases/2010/02/100201171641.htm.
12. Hans Pretzsch, et al., "Forest stand growth dynamics in Central Europe have accelerated since 1870," *Nature Communications* 5 (4967) 2014. https://www.nature.com/articles/ncomms5967.

pushed to the brink of ecological collapse." The latter statement is much more likely to be printed in newspapers and broadcast on worldwide media. It is not objective science, it's instead a sensationalist prediction that has no basis in science or logic. What is required here is an appeal for critical thinking among the populace in order to distinguish between the factual and the predictive, and between sober language and apocalyptic pronouncements with no foundation.

The Historical Record of CO_2 and Temperature in the Atmosphere

All the carbon dioxide in the atmosphere originally came from inside the Earth. During the early life of the planet, the Earth was much hotter, and there was much more volcanic activity than there is today. The heat of the Earth's core caused carbon and oxygen to combine to form CO_2, which became a significant part of the Earth's early atmosphere. It was perhaps one of the most abundant components until photosynthesis and eventually calcification evolved. Most of the carbon dioxide in the oceans comes from the atmosphere, although some was injected directly from ocean vents. It is widely accepted that the concentration of CO_2 was far higher in the Earth's atmosphere before modern-day life forms evolved during the Cambrian Period, which began 570 million years ago. It was also during that time that a number of marine species evolved the ability to control calcification, an example of the more-general term "biomineralization." This allowed these species to build hard shells of calcium carbonate ($CaCO_3$) around their soft bodies, thus providing a type of armor plating. Early shellfish, such as clams, arose more than 500 million years ago, when atmospheric CO_2 was 10 to 15 times higher than it is today.[13] Clearly, the pH level of the oceans then did not cause the extinction of all corals or shellfish or they would certainly not be here today. Why, then, are we told that even with today's much lower carbon dioxide level, that it is causing damage to calcifying species?

The most common argument is along the lines of "today's species of corals and shellfish are not adapted to the level of CO_2 that ancient species were familiar with. Acidification is happening so quickly that species will not be able to adapt to higher levels of CO_2." This is a nonsensical argument in that from a biochemical perspective there is no reason to believe these species have lost their ability to calcify at the

---

13. "The Cambrian Period (544-505 mya)," Virtual Fossil Museum. http://www.fossilmuseum.net/Paleobiology/Paleozoic_paleobiology.htm#Cambrian.

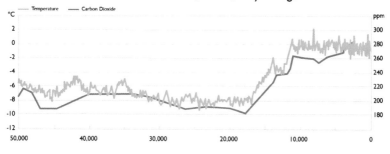

Figure 87. Reconstructions of air temperature and CO_2 concentration from Vostok Station, Antarctica, 50,000 to 2,500 years ago. Carbon dioxide concentration fell to a little above 180 ppm 18,000 years ago; only 30 ppm above the level that causes plants to die for lack of CO_2.*

* Joanne Nova, "The 800-year lag in CO_2 after temperature – graphed," JoNova, August 18, 2013. http://joannenova.com.au/global-warming-2/ice-core-graph/.

higher carbon dioxide levels, such as the ones that existed for millions of years in the past. The ancestors of every species alive today survived through millennia during which conditions sometimes changed very rapidly, such as when an asteroid caused the extinction of dinosaurs and many other species 65 million years ago. While many more species have become extinct than are alive today, it must be said that the species that survived those times have proven the most resilient of all.

Not many people stop to think that every individual of every species on Earth today represents a continuously successful line of reproduction from the beginning of life.

Geological Timescale: Concentration of CO_2 and Temperature Fluctuations

As far as is known, there was only one other period in the Earth's history when carbon dioxide was nearly as low as it has been during the past 2.5 million years of the Pleistocene Ice Age. During the late Carboniferous Period and into the Permian and Triassic Periods, carbon dioxide was drawn down from about 4,000 ppm to about 400 ppm, probably owing to the advent of vast areas of forest that pulled CO_2 out of the atmosphere and incorporated it into wood and thus into coal. We know from Antarctic ice cores that carbon dioxide was drawn down to as low as 180 ppm during the Pleistocene at the peak of the most recent glacial advance (see Fig. 87). We know with considerable confidence that this is the lowest it has been since the beginning of life on Earth. This level of CO_2 is only 30 ppm above the threshold for the survival of plants. These periods of low atmospheric carbon dioxide, as is still the case presently, are the exception to the much longer periods when CO_2 was more than 1,000 ppm, and often much higher. For this reason alone, the possibility that present and future atmospheric CO_2 levels will cause significant harm to calcifying marine life should be rejected. However, there are also a number of other factors that bring the ocean acidification hypothesis into question.

The Adaptation of Species to Changing Environmental Conditions

People have a tendency to assume that it takes thousands or millions of years for species to adapt to changes in the environment. This is not the case. Even species with relatively long periods between reproduction can adapt relatively quickly when challenged by rapidly changing environmental conditions. In fact, it is rapidly changing environmental conditions that foster rapid evolutionary change and adaptation.[14] Stephen Jay Gould explains it well in his classic book *Wonderful Life*, which focuses on the Burgess Shale fossils from the Cambrian Explosion and the evolution of vast numbers of species beginning 544 million years ago.[15]

Most of the invertebrates that have developed the ability to produce calcium carbonate armor are also capable of relatively rapid adaptation to changes in their environment due to two distinct factors. Firstly, they reproduce at least annually and sometimes more frequently. This means their progeny are tested on an annual basis for suitability to a changing environment. And secondly, these species produce hundreds to thousands of offspring every time they reproduce. This greatly increases the chance that genetic mutations that are better suited to changes in environmental conditions will occur in some offspring.

A number of studies have demonstrated that changes in an organism's genetic make-up, or genotype, is not the only factor that allows species to adapt to changing environmental conditions. Many marine species inhabit coastal waters for some or all of their lives where they are exposed to much wider ranges of pH, CO_2, O_2, temperature,

---

14. Jeroen Boeye, et al., "More rapid climate change promotes evolutionary rescue through selection for increased dispersal distance," *Evolutionary Applications* 6, no. 2 pp353-364, February 6, 2013. https://www.ncbi.nlm.nih.gov/pmc/articles/PMC3586623/.
15. Stephen Jay Gould, *Wonderful Life: The Burgess Shale and the Nature of History*, W.W. Norton and Company, 1989. https://www.amazon.com/Wonderful-Life-Burgess-Nature-History/dp/039330700X.

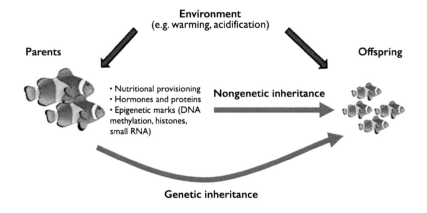

Figure 88. Phenotypic plasticity (environmental) and trans-generational plasticity (non-genetic inheritance) are examples of adaptations that are not directly related to changes in the DNA of an organism. These forms of adaptation can occur during the lifetime of an individual or in a single generation.*

* Philip L. Munday, "Transgenerational acclimation of fishes to climate change and ocean acidification," F1000Prime Rep, November 4, 2014. https://www.ncbi.nlm.nih.gov/pmc/articles/PMC4229724/.

and salinity than occur in the open ocean. Two distinct physiological mechanisms exist whereby adaptation to environmental change can occur much more rapidly than by change in the genotype through genetic evolution (see Fig. 88).

The first of these is phenotypic plasticity, which is the ability of one genotype to produce more than one phenotype[16] when exposed to different environments.[17] In other words, a specific genotype can express itself differently due to an ability to respond in different ways to variations in environmental factors. This helps to explain how individuals of the same species with nearly identical genotypes can successfully inhabit very different environments. Examples of this in humans are the ability to acclimatize to different temperature regimes and different altitudes. There is no change in the genotype, but there are changes in physiology.

---

16. In biology, a genotype is the composition of the DNA in an organism's genes. A phenotype is the expression of those genes in the physical organism itself. Differences in environmental conditions can result in different expressions of the identical genotype in the phenotype. This is one aspect of the expression, "is it nature or is it nurture?" In fact, it is a combination of the two, and one of the more interesting subjects in biology and psychology.
17. Trevor D. Price, et al., "The role of phenotypic plasticity in driving genetic evolution," Proceedings of the Royal Society B: Biological Sciences 270 (1523) pp1433-1440. https://www.ncbi.nlm.nih.gov/pmc/articles/PMC1691402/.

The second and more fascinating factor is trans-generational plasticity, which is the ability of parents to pass their adaptations to their offspring.[18] Another paper pointed out that:

Contemporary coastal organisms already experience a wide range of pH and CO_2 conditions, most of which are not predicted to occur in the open ocean for hundreds of years – if ever.[19]

The authors used what they called:

A novel experimental approach that combined bi-weekly sampling of a wild, spawning fish population (Atlantic silverside, Menidia menidia) with standardized offspring CO_2 exposure experiments and parallel pH monitoring of a coastal ecosystem.

The parents and offspring were exposed to CO_2 levels of 1,200 ppm and 2,300 ppm compared with today's ambient level of 400 ppm. The scientists report that:

Early in the season (April), high CO_2 levels significantly…reduced fish survival by 54 percent (2012) and 33 percent (2013) and reduced 1- to 10-day post-hatch growth by 17 percent relative to ambient conditions.

However, they found that:

Offspring from parents collected later in the season became increasingly CO_2-tolerant until, by mid-May, off-spring survival was equally high at all CO_2 levels."

This indicates that a coastal species of fish is capable of adapting to high levels of carbon dioxide in a very short time. It also indicates that this same species would not even notice the relatively slow rate at which carbon dioxide is increasing in the oceans today, if it is indeed increasing at all.

The changes that have occurred to the Earth's climate over the past 300 years since the depth of the Little Ice Age in about 1700 are in

---

18. Eva Jablonka and Gal Raz, "Transgenerational Epigenetic Inheritance: Prevalence, Mechanisms, and Implications for the Study of Heredity and Evolution," PubMed, June 2009. https://pubmed.ncbi.nlm.nih.gov/19606595/.
19. Christopher S. Murray, et al., "Offspring sensitivity to ocean acidification changes seasonally in a coastal marine fish," *Marine Ecology Progress Series* 504 pp1-11, May 14, 2014. https://www.int-res.com/articles/feature/m504p001.pdf.

no way unusual or unique in history. During the past 3,000 years, but a blink in geological time, there has been a succession of warm periods and cool periods. There is no record of species extinction due to climatic change during these periods. And during the longer time periods such as from 500 million years ago, the biodiversity of life has increased dramatically despite five major extinction events. Nothing that is happening today comes even close to the changes that have occurred through life's long history.

The Buffering Capacity of Seawater

Over the millennia, the oceans have received minerals dissolved in rainwater from the land. Most of these are in the form of ions such as chloride, sodium, sulphate, magnesium, potassium, and calcium. Underwater hydrothermal vents and submarine volcanoes also contribute to the oceans' salt content. These elements have come to make up about 3.5 percent of seawater by mass, thus giving seawater some unique properties compared with fresh water. The salt content of the sea has been relatively stable for hundreds of millions, even billions, of years, as mineralization on the sea floor balances out the new salts entering the sea.[20]

Seawater has a powerful buffering capacity due to the salts and carbon dioxide dissolved in it. Buffering capacity is the ability of liquids to resist change in pH when an acidic or basic compound is added to the liquid. For example, one micromole of hydrochloric acid added to one kilo of distilled water at pH 7.0 (neutral) causes the pH to drop to nearly 6.0. If the same amount of hydrochloric acid is added to seawater at pH 7, the resulting pH is 6.997 – a change of only 0.003 of a pH unit. Thus, seawater has approximately 330 times the buffering capacity of freshwater.[21] In addition to the buffering capacity, there is also another factor. The Revelle factor is named after Roger Revelle, the former director of the Scripps Institute of Oceanography. The Revelle factor determines that if atmospheric carbon dioxide is doubled, the dissolved CO_2 in the ocean will only rise by 10 percent.[22] This is

---

20. Wallace S. Broecker, "Chemical Oceanography," Harcourt, Brace, Jovanovich Inc., 1974. https://www.ldeo.columbia.edu/~broecker/Chemical%20Oceanography.pdf.
21. R. E. Zeebe, et al., "Carbon Dioxide, Dissolved (Ocean)," *Encyclopedia of Paleoclimatology and Ancient Environments*, ed. V. Gornitz, Kluwer Academic Publishers, *Earth Science Series*, 2008. https://www.soest.hawaii.edu/oceanography/faculty/zeebe_files/Publications/ZeebeWolfEnclp07.pdf.
22. Ibid.

something that is not being recognized by the people claiming that rising carbon dioxide will cause extinctions in the oceans.

CO_2 is not only the basis for all life, it is also why the oceans have a pH that is not so alkaline that they could not support life at all. To get a buffered solution, you have to combine a weak acid with a strong base (or a strong acid with a weak base). In the case of the oceans you can think of the strong base as a combination of sodium hydroxide (NaOH) – which is the main constituent of the drain cleaner, Draino – calcium hydroxide ($Ca(OH)_2$), and magnesium hydroxide ($Mg(OH)_2$). Without the carbon dioxide dissolved in seawater, the oceanic pH would be at about 11.3, similar to household ammonia. Seawater with a pH of 11.3 would not support life. It is only because of the CO_2 dissolved in seawater that the pH is a mildly alkaline 7.5 to 8.4. The fact is, it is carbon dioxide itself that is the main reason seawater has such a high buffering capacity. It is a weak acid and the salts are mainly strong bases. Once again carbon dioxide proves to be an essential ingredient for the evolution of life on Earth, which began in the sea.

It is widely stated in the literature that the pH of the oceans was 8.2 before industrialization (1750) and that, owing to human CO_2 emissions, it has since dropped to 8.1.[23,24] The fact is, no one measured the pH of ocean water in 1750. The concept of pH was not conceived of until 1909, and an accurate pH meter was not available until 1924. The assertion that more than 250 years ago ocean pH was 8.2 is an unsubstantiated guess rather than any kind of actual measurement. Measuring pH accurately in the field, to 0.1 of a pH unit is not a simple procedure even today. In addition, there is no global-scale monitoring of the pH of the oceans. This is largely because genuine oceanographers know about the overwhelming buffering capacity of seawater, so they do not expect the global acid-base balance in the oceans to change substantially.

The predictions of change in ocean pH owing to carbon dioxide in the future are based on the same assumptions that resulted in the estimate of pH 8.2 in 1750 when we of course had no measurement of the pH of the oceans at that time. By simply extrapolating from the claim that pH has dropped from 8.2 to 8.1 during the past 265 years,

---

23. R. E. Zeebe, "History of Seawater Carbonate Chemistry, Atmospheric CO_2, and Ocean Acidification," *Annual Review of Earth and Planetary Science* 40, pp141-165, 2012. https://www.soest.hawaii.edu/oceanography/faculty/zeebe_files/Publications/ZeebeAR12.pdf.
24. "Ocean Acidification in the Pacific Northwest," NOAA, May 2014. https://wsg.washington.edu/wordpress/wp-content/uploads/OA18PNWFacts14V5.pdf.

the models calculate that pH will drop by 0.3 of a pH by the year 2100. And, of course, CO_2 is claimed to be the culprit once again.

Many scientists have repeated the claim that the ocean's pH has dropped by 0.1 during the past 265 years. They should be challenged to provide data or proxies from 1750 that supports their inference. Observations from three eminent oceanographers, including Harald Sverdrup, former director of the Scripps Institute of Oceanography, stated in a book covering all aspects of ocean physics, chemistry, and biology bring into question the alarmist scientists' assertion. The book was written before the subject of climate change and carbon dioxide became politicized.

> *Sea water is a very favorable medium for the development of photosynthetic organisms. It not only contains an abundant supply of CO_2, but removal or addition of considerable amounts results in no marked changes of the partial pressure of CO_2 and the pH of the solution, both of which are properties of importance in the biological environment…If a small quantity of a strong acid or base is added to pure water, there are tremendous changes in the numbers of H+ and OH– ions present, but the changes are small if the acid or base is added to a solution containing a weak acid and its salts or a weak base and its salts. This repression of the change in pH is known as buffer action, and such solutions are called buffer solutions. Sea water contains carbonic and boric acids and their salts and is, therefore, a buffer solution. Let us consider only the carbonate system. Carbonate and bicarbonate salts of strong bases, such as occur in sea water, tend to hydrolyze, and there are always both H+ and OH– ions in the solution. If an acid is added, carbonate is converted to bicarbonate and the bicarbonate to carbonic acid, but, as the latter is a weak acid (only slightly dissociated), relatively few additional hydrogen ions are set free. Similarly, if a strong base is added, the amount of carbonate increases, but the OH– ions formed in the hydrolysis of the carbonate increase only slightly. The buffering effect is greatest when the hydrogen ion concentration is equal to the dissociation constant of the weak acid or base – that is, when the concentration of the acid is equal to that of its salt.[25]*

In addition to this, a study has also been published in which the pH of the oceans was reconstructed from 1708 to 1988, based on the boron isotopic composition of a long-lived massive coral from Flinders Reef in the western Coral Sea of the southwestern Pacific.[26] The report

---

25. H. U. Sverdrup, et al., *The Oceans, Their Physics, Chemistry, and General Biology*, Prentice-Hall: New York, 1942. https://publishing.cdlib.org/ucpressebooks/view?docId=kt167nb66r.
26. C. E. Pelejero, E. Calvo, M. T. McCulloch, J. F. Marshall, et al., "Preindustrial to Modern Interdecadal

concluded that there was no notable trend toward lower isotopic values over the 300-year period investigated. This indicates that there has been no change in ocean pH over that period in the western Coral Sea region. This study, in which actual measurements of a reliable proxy were made, is much more credible and reliable than a guess based on assumptions that have not even been tested.

In many ways, the assertions made about the degree of pH change caused by a given level of atmospheric carbon dioxide are analogous to the claims made about the degree of atmospheric temperature rise that might be caused by a given level of atmospheric carbon dioxide. This is termed "sensitivity" and the literature becomes very confusing when the subject of sensitivity is researched. Perhaps the assumptions used to estimate future ocean pH are as questionable as those used to estimate the increase in temperature from increases in atmospheric CO_2 (see Fig. 89).

The most serious problem with the assertion that pH has dropped from 8.2 to 8.1 since 1750 is that there is no universal pH in the world's oceans. The pH of the oceans varies far more than 0.1 on a daily, monthly, annual, and geographic basis. In the offshore oceans, pH typically varies geographically from 7.5 to 8.4, or 0.9 of a pH unit. A study in offshore California shows that pH can vary by 1.43 of a pH unit on a monthly basis.[27] This is nearly five times the change in pH that computer models forecast during the next 80 years to 2100. In coastal areas that are influenced by run-off from the land, pH can be as low as 6.0 and as high as 9.0.[28]

The Humboldt Current is a large area of upwelling ocean water off the coasts of Chile and Peru. It has some of the lowest pH levels found naturally in the offshore oceans. Here the pH of the seawater is 7.7 to 7.8. If the ocean average pH is now 8.1, the water in the Humboldt Current is already at a slightly lower pH than is predicted by 2100. Upwelling waters tend to be lower in pH than other areas of the ocean; this is because of two factors. First, the water has been at a depth where the remains of sea creatures fall down and decompose into nutrients and carbon dioxide; this tends to drive pH levels down. Second, the

---

Variability in Coral Reef pH," *Science* 309, pp2204-2207, 2005. https://science.sciencemag.org/content/309/5744/2204.figures-only.

27. US House Committee on Natural Resources, "CEQ Draft Guidance for GHG Emissions and the Effects of Climate Change," Testimony of Professor John R. Christy, University of Alabama in Huntsville, May 13, 2015. https://docs.house.gov/meetings/II/II00/20150513/103524/HHRG-114-II00-Wstate-ChristyJ-20150513.pdf.

28. G. E. Hofmann, et al., "High-Frequency Dynamics of Ocean pH: A Multi-Ecosystem Comparison," *PLoS ONE* 6, no. 12, December 9, 2011. https://journals.plos.org/plosone/article?id=10.1371/journal.pone.0028983.

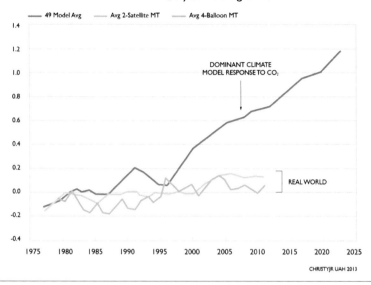

Figure 89. Here is a comparison of the predictions made by computer models with real observations supported by weather balloons and satellites of mid-tropospheric temperature. The reason they got it right before 1995 is that the modelers knew that temperature history when the models were made, so they adjusted the models to follow the historic trends. The future was not so cooperative. This is called "models running hot."*

* Richard S. Lindzen, PhD, "On Climate Sensitivity," CO_2 Coalition, April 8, 2020. http://co2coalition.org/publications/on-climate-sensitivity/.

water that is upwelling to form the Humboldt Current is water that downwelled (sank) around Antarctica, and being cold, it had a high solubility for CO_2 at the ocean-atmosphere interface. Ocean water that sinks at the poles eventually comes to the surface where it is warmed. As the water warms, there is a natural outgassing of some of the carbon dioxide that was absorbed in the Antarctic (see Fig. 90, page 188).

Despite – actually because of – its low pH, the upwelling waters of the Humboldt Current produce 20 percent of the world's wild fish catch, consisting largely of anchovies, sardines, and mackerel.[29] The Humboldt Current waters produce the world's highest catch per unit area as well. The proponents of ocean acidification don't seem to realize that carbon dioxide is the food for plankton in the sea, just like it is for plants on the land. Higher CO_2 levels are causing a greening of the oceans just as it is on the land.

---

29. Wikipedia, "Humboldt Current," September 16, 2020. https://en.wikipedia.org/wiki/Humboldt_Current.

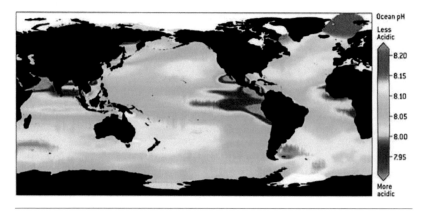

Figure 90. World map depicting the pH of the oceans, including the large area of lower pH seawater off the west coast of South America. To be scientifically correct, the scale of ocean pH on the right should read "More Basic" at the top and "Less Basic" at the bottom. From *Scientific American*, March 2006* This journal has "gone tabloid" too.

* S. C. Doney, "The Dangers of Ocean Acidification," *Scientific American*, March.2006. https://www.scientificamerican.com/article/the-dangers-of-ocean-acid/.

The basis for the food chain in Humboldt Current includes large blooms of coccolithophores, a calcifying phytoplankton that produces symmetrical calcium carbonate shields to protect itself from predators. The White Cliffs of Dover are composed of the shells of coccolithophores. To quote from one of the more thoroughly researched papers on the subject:

> *These biome-specific pH signatures disclose current levels of exposure to both high and low dissolved CO_2, often demonstrating that resident organisms are already experiencing pH regimes that are not predicted until 2100.*

The authors continue, by asserting, "The effect of Ocean Acidification (OA) on marine biota is quasi-predictable at best."[30] It is refreshing to read an opinion that is not so certain about predicting the future of an ecosystem as complex as the world's oceans.

Scientists working at oceanographic institutes in the United Kingdom and Germany published a paper in 2015 that explored the possibility that the asteroid that struck the Earth 65 million years ago caused ocean acidification. Along with the extinction of terrestrial and marine dinosaurs, 100 percent of ammonites and 90 percent of coccolithophores, both calcifying species, became extinct. The study

---

30. E. Hofmann, et al., "High-Frequency Dynamics of Ocean pH: A Multi-Ecosystem Comparison," *PLoS ONE* 6, no. 12, December 19, 2011. https://journals.plos.org/plosone/article?id=10.1371/journal.pone.0028983.

considered the possibility that 6,500 Gt (billion tons) of carbon as CO_2 were produced by the vaporization of carbonaceous rock and wildfires because of the impact. The authors concluded, "Our results suggest that acidification was most probably not the cause of the extinctions."[31]

Six thousand five hundred Gt is the equivalent of 650 years of carbon dioxide emissions at the current global rate of about 10 Gt carbon as CO_2 per year. Given that atmospheric carbon dioxide concentration was about 1,000 ppm at the time of the impact, the addition of 6,500 Gt of carbon as CO_2 would have raised the concentration to approximately 4,170, which is about 10 times higher than in 2015 and about five times higher than it may be in 2100

The Ability of Calcifying Species to Control the Biochemistry at the Site of Calcification

All organisms are able to control the chemistry of their internal organs and biochemical processes. The term "homeostasis" means that an organism can maintain a desirable state of chemistry, temperature, pH, etc. within itself under a range of external conditions.[32]

One of the most important aspects of homeostasis is "osmoregulation." There are two biological strategies for accomplishing osmoregulation in marine organisms. The osmoregulators, which include most fishes, maintain their internal salinity at a different level from their environment. It requires energy to counteract the natural osmotic pressure that tends to equalize an organism's internal salinity with the salinity of the water it inhabits. The osmoconformers, which include most of the invertebrate species, maintain their salt content at the same osmotic pressure as their environment. Both osmoregulators and osmoconformers alter the ratio of the salts inside themselves compared with the ratio of salts in the water they inhabit.[33]

The osmoregulators are best illustrated by the three examples of freshwater fish, saltwater fish, and fish that are able to live in both freshwater and saltwater. Freshwater fish must be able to retain salts in their bodies and they are able to repel and expel fresh water and they can recover salts from their kidneys before excretion. Saltwater fish

---

31. Toby Tyrrell, "Severity of ocean acidification following the end-Cretaceous asteroid impact," *PNAS* 112, no. 21 pp6556-61. https://www.pnas.org/content/112/21/6556.
32. J. M. Wood, "Bacterial Osmoregulation: A Paradigm for the Study of Cellular Homeostasis," *Annual Review of Microbiology* 65 pp215-238, October 2011. https://www.annualreviews.org/doi/abs/10.1146/annurev-micro-090110-102815.
33. Kenneth S. Saladin, "Osmoregulation," Biology Encyclopedia Forum. http://www.biologyreference.com/Oc-Ph/Osmoregulation.html.

are able to retain water while excreting salts through their gills, thus lowering their internal salt content compared to seawater. Fish such as salmon and eels that spend part of their lives in fresh water and part in saltwater are able to transform their bodily functions as they move from one environment to the other, an astonishing feat.[34] This ability to radically alter internal biochemistry, for some species, numerous times during their lifetime, is a classic case of phenotypic plasticity. The genes do not change, but the instructions they give to the organism change in concert with the changes in their environment.

Some osmoconformers, such as starfish and sea urchins, can only tolerate a narrow range of external salinity while others, such as mussels and clams – that can isolate themselves from the environment by closing their shells – can tolerate a wide range of external salinity.[35] There is a good reason why many species that live near river estuaries and in intertidal zones, such as clams, mussels, oysters, and barnacles have shells that can close tightly. These environments typically undergo large and frequent variations in salinity, pH, sediment, oxygen concentration, and temperature. Species that live in intertidal zones must be capable of surviving out of the water and in the atmosphere for varying periods of time.

Osmoregulation is a good example of how species are able to adapt to environments that would otherwise be hostile to life. Controlled calcification is another biological function that depends on species' ability to alter and control their internal chemistry.

The ocean acidification narrative is based almost entirely on the chemistry of seawater and the chemistry of calcium and carbon dioxide. It is true that the shell of a dead organism will gradually dissolve in water with a lowered pH;[36] however, it cannot be inferred directly from this that the shell, or carapace, or coral structure of a species will dissolve under similar pH while the organism is alive. Even if some dissolution is occurring, as long as the organism builds calcium carbonate faster than it dissolves, the shell will grow. If this were not the case, it would be impossible for the duck mussel, Anodonta anatina, to survive in a laboratory experiment at pH 3.0 for 10 days without significant shell loss.[37]

---

34. "Surviving in Salt Water, American Museum of Natural History. https://www.amnh.org/exhibitions/water-h2o-life/life-in-water/surviving-in-salt-water.
35. Marion McClary, "Osmoconformer," *Encyclopedia of the Earth*, June 20, 2014. http://www.eoearth.org/view/article/155074/.
36. "What is Ocean Acidification?" National Oceanic and Atmospheric Administration, PMEL Carbon Program. https://www.pmel.noaa.gov/co2/story/What+is+Ocean+Acidification%3F.
37. T. P. Mäkelä, et al., "The Effects of Low Water pH on the Ionic Balance in Freshwater Mussel Anodonta

This is an extreme example, as it is outside natural conditions. It is, however, a proven fact that calcification occurs in freshwater species of mussels and clams at pH 6.0, which is well into the range of genuine acidity. The Louisiana pearlshell, Margaritifera hembeli, is actually restricted to waters with a pH of 6.0 to 6.9. In other words, it requires acidic water to survive.[38] This does not mean that all marine species that calcify will tolerate pH 6.0, it only shows that there are organisms that can calcify at much lower pH than is found in ocean waters today or that are not possible in seawater even under extreme scenarios.

The coccolithophores account for about 50 percent of all calcium carbonate production in the open oceans (not including coral reefs and other stationary calcifiers). A laboratory study found that:

> *The coccolithophore species Emiliania huxleyi are significantly increased by high CO_2 partial pressures* and that *over the past 220 years there has been a 40 percent increase in average coccolith mass* and that **in a scenario where the CO_2 in the world's oceans increases to 750 ppm, coccolithophores will double their rate of calcification and photosynthesis.**[39] (My emphasis.)

This is good news for the ocean's primary production and fisheries production up the food chain. It demonstrates that higher levels of carbon dioxide will not only increase productivity in plants, both terrestrial and aquatic, but will also boost the productivity of one – if not the most important – of the calcifying species in the oceans.

The reason that calcifying marine organisms can calcify under a wider range of pH values than one would expect from a simple chemical calculation is that they can control their internal chemistry at the site of calcification. The proponents of dangerous ocean acidification do not consider this. If the internal biology of organisms were strictly determined by the chemical environment around them, it is unlikely there would be any life on Earth. This is why there are cell membranes, cell walls, and shells; to control what goes in and out of them in order to maintain an optimum biochemical environment for functions necessary for survival and growth.

---

anatina L," *Annales Zoologici Fennici* 29, pp169-175, December 1992. http://www.sekj.org/PDF/anzf29/anz29-169-175.pdf.
38. Wendell R. Haag, *North American Freshwater Mussels: Natural History, Ecology, and Conservation*, Cambridge University Press, 2012. http://www.langtoninfo.com/web_content/9780521199384_frontmatter.pdf.
39. M. D. Iglesias-Rodriguez, et al., "Phytoplankton Calcification in a High-CO_2 World," *Science* 320, 5847 pp336-340, April 2008. http://www.sciencemag.org/content/320/5874/336.full#F1.

As mentioned earlier, it was at the beginning of the Cambrian Period approximately 540 million years ago that marine species of invertebrates evolved the ability to control the crystallization of calcium carbonate as an armor plating to protect themselves from predators. It is hypothesized that this ability stemmed from a long-standing previous ability to prevent spontaneous calcium carbonate crystallization to protect essential metabolic processes. Surprisingly, the common denominator in the anti-calcification to calcification history is mucus, often referred to as "slime," the kind of slime seen in the trail of a slug or land-snail.[40]

The abstract from the paper cited just previously sums up this hypothesis well:

> *The sudden appearance of calcified skeletons among many different invertebrate taxa at the Precambrian-Cambrian transition may have required minor reorganization of pre-existing secretory functions. In particular, features of the skeletal organic matrix responsible for regulating crystal growth by inhibition may be derived from mucus epithelial excretions. The latter would have prevented spontaneous calcium carbonate overcrusting of soft tissues exposed to the highly supersaturated Late Proterozoic ocean...a putative function for which we propose the term 'anticalcification.' We tested this hypothesis by comparing the serological properties of skeletal water-soluble matrices and mucus excretions of three invertebrates – the scleractinian coral Galax eafascicularis and the bivalve molluscs Mytilus edulis and Mercenaria mercenaria. Crossreactivities recorded between muci and skeletal water-soluble matrices suggest that these different secretory products have a high degree of homology. Furthermore, freshly extracted muci of Mytilus were found to inhibit calcium carbonate precipitation in solution.*[41]

The authors found that the muci produced by a coral, a mussel, and a clam were chemically very similar, indicating inheritance from a common ancestor earlier in the Precambrian. The mucus produced by invertebrates has a number of known functions. It assists with mobility, acts as a barrier to disease and predators, helps with feeding, acts as a homing device, and prevents desiccation.[42] Perhaps you have seen

---

40. F. Marin, et al., "Skeletal matrices, muci, and the origin of invertebrate calcification," Proceedings of the National Academy of Sciences of the United States of America 93, no. 4 pp1554-1559, February 20, 1996. https://www.pnas.org/content/93/4/1554.abstract?tab=related.
41. Ibid.
42. M. W. Denny, "Invertebrate mucous secretions: functional alternatives to vertebrate paradigms," Symposia of the Society for Experimental Biology 43 pp337-366, December 31, 1988. https://europepmc.org/article/med/2701483.

the egg case of a moon snail on the beach at a very low tide. The case, which contains the tiny eggs, is made with sand glued together with the snail's mucus (see Fig. 91).

The authors postulate that the mucus is also central in the calcification process. This explains how the chemistry at the site of calcification can be isolated from the chemistry of the seawater. Calcification can occur in and under the mucus layer where the organism can control the chemistry. The "periostracum" is the leathery proteinaceous outermost layer on many shells:

> *The formation of a shell requires certain biological machinery. The shell is deposited within a small compartment, the extrapallial space, which is sealed from the environment by the periostracum, a leathery outer layer around the rim of the shell, where growth occurs. This caps off the extrapallial space, which is bounded on its other surfaces by the existing shell and the mantle. The periostracum acts as a framework from which the outer layer of carbonate can be suspended, but also, in sealing the compartment, allows the accumulation of ions in concentrations sufficient for crystallization to occur. The accumulation of ions is driven by ion pumps packed within the calcifying epithelium. The organic matrix forms the scaffold that directs crystallization,*

Figure 91. A mother moon snail with its unique egg case. The case is made with sand glued together with the snail's mucus. The eggs are embedded in the case. This is only one of the many uses for mucus among calcifying species.

and the deposition and rate of crystals is also controlled by hormones produced by the mollusk.[43]

The quotation above makes it clear that calcifiers have a high degree of sophistication in controlling the calcification process. The clear implication is that calcification can be successfully achieved despite a varying range of environmental conditions that would interfere with or end the process if it were not controlled. This does not appear to have been considered by most, if not all, of the authors who propose that ocean acidification will exterminate a large portion of calcifying species within a few decades (see Fig. 92).

Much of the concern about ocean acidification in the literature focuses on carbonate chemistry. When the pH of seawater lowers, the bicarbonate ion (HCO_3) becomes more abundant while the carbonate ion (CO_3) becomes less abundant. This is predicted to make it more difficult for calcifying species to obtain the CO_3 required for calcification. It does not appear to be considered that the calcifying species may be capable of converting HCO_3 to CO_3 internally.

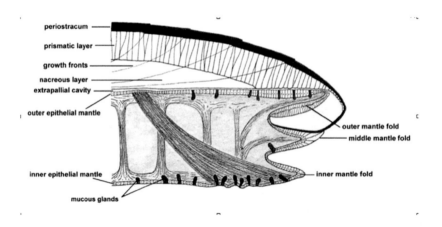

Figure 92. Graphic of the structures in a bivalve mollusk, such as a clam. The periostracum seals the extrapallial cavity where the calcification takes place, building new shell. The process is not done in the seawater; it is done internally where the biochemistry can be controlled for optimal efficiency.

---

43. "Periostracum," *Encyclopaedia Britannica*, 2015. http://www.britannica.com/science/periostracum.

There are very few references to journal articles after 1996 that investigate the biochemical processes involved in calcification. There is clearly much more to be learned about this complex process that is practiced by so many species. The paper cited previously by Marin, et al., is the most thorough investigation and discussion of the subject found. Yet, there are hundreds, if not thousands of articles that predict dire consequences from ocean acidification during this century. A search of "ocean acidification" on the internet delivers 8,650,000 results.

A recent study published in the Proceedings of the National Academy of Sciences highlights how resilient coral reefs are to changes in ocean pH. A five-year study of the Bermuda coral reef shows that during spurts in growth and calcification the seawater around the reef undergoes a rapid reduction in pH.[44] This reduction in pH is clearly not causing a negative reaction from the reef, as it is associated with rapid growth. The study found that the reason the pH dropped during growth spurts is because of the CO_2 emitted by the reef due to increased respiration. It was determined that the growth spurts were the result of offshore blooms of phytoplankton drifting into the reef and providing an abundant food supply for the reef polyps. The clear conclusion from the study is that coral growth can increase even though the growth itself results in a reduction in pH in the surrounding seawater. A summary of the study in *New Scientist* concluded:

> *These corals didn't seem to mind the fluctuations in local acidity that they created, which were much bigger than those we expect to see from climate change. This may mean that corals are well equipped to deal with the lower pH levels. It follows from the discussion above that this is likely due to the fact that the coral polyps can control their own internal pH despite the decrease in pH in their environment.*[45]

A Warmer Ocean May Emit CO_2 Back into the Atmosphere

While today's atmosphere contains about 850 Gt of carbon as CO_2, the oceans contain about 38,000 Gt of carbon – nearly 45 times as much as the atmosphere contains. The ocean either absorbs or emits carbon

---

44. K. L. Yeakel, "Shifts in coral reef biogeochemistry and resulting acidification linked to offshore productivity," *PNAS* 2015: 507021112v1-201507021. https://www.pnas.org/content/early/2015/11/04/1507021112.
45. M. Slezak, "Growing corals turn water more acidic without suffering damage," *New Scientist*, November 13, 2015. https://www.newscientist.com/article/dn28468-growing-corals-turn-water-more-acidic-without-suffering-damage/.

dioxide at the ocean-atmosphere interface, depending on the CO_2 concentrations in the atmosphere and the sea below, and in the salinity and temperature of the sea. At the poles, where seawater is coldest and densest and has the highest solubility for carbon dioxide, seawater sinks into the abyss, taking the carbon dioxide down with it. In regions of deep seawater upwelling such as off the coasts of Peru, California, West Africa, and the northern India Ocean, seawater rich in CO_2 fertilizes plankton blooms that feed great fisheries. The phytoplankton near the surface consumes some of the carbon dioxide, and some is outgassed to the atmosphere.

As mentioned above, we do not have the ability to determine how much carbon dioxide is absorbed by the oceans, how much is outgassed back into the atmosphere, or the net effect of these phenomena. What we do know is that if the oceans warm, as the proponents of human-caused global warming say they will, the oceans will tend to release CO_2 into the atmosphere because warm seawater at 30°C can dissolve only about half as much carbon dioxide as cold seawater at 0°C does. This will be balanced against the tendency of increased atmospheric carbon dioxide to result in more absorption of CO_2 by the oceans. It does not appear as though anyone has done the calculation of the net effect of these two competing factors under varying circumstances, or if such a calculation is even possible.

Experimental Results on Effects of Reduced pH on Calcifying Species

In his thorough and inclusive analysis of peer-reviewed experimental results on the effect of reduced pH on five factors (calcification, metabolism, growth, fertility, and survival) among marine calcifying species, Craig Idso of the CO_2science.com website, provides a surprising insight. Beginning with 1,103 results from a wide range of studies, the results are narrowed down to those within a 0.0 to 0.3 reduction in pH units.[46] A review of these many studies, all of which use direct observation of measured parameters, indicates that the overall predicted effect of increased carbon dioxide on marine species would be positive rather than negative (see Fig. 93). This further reinforces the fact that carbon dioxide is essential for life, that CO_2 is at a historical low concentration

---

46. "Ocean Acidification Database," CO_2 Science, 2015. http://www.co2science.org/data/acidification/results.php. See also http://www.co2science.org/subject/subject.php.

during this Pleistocene Ice Age, and that more CO_2 rather than less would be generally beneficial to life on Earth.

Conclusion

There is no solid evidence that ocean acidification is the dire threat to marine species that many researchers have claimed. The entire premise is based upon an assumption of what the average pH of the oceans was 265 years ago when it was not even possible to measure pH anywhere at that time, never mind over all the world's oceans. Laboratory experiments in which pH was kept within a range that proponents of ocean acidification contend may occur during this century show a slight positive effect on five critical factors: calcification, metabolism, growth, fertility, and survival. Of most importance is the fact that those raising the alarm about ocean acidification do not take into account the ability of living species to adapt to a range of environmental conditions. This is one of the fundamental characteristics of evolution and of life itself.

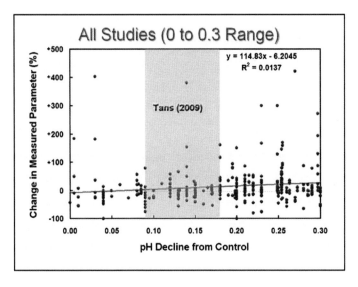

Figure 93. All peer-reviewed experimental results for a pH decrease of 0.0 to 0.3 from the present value. (Prediction of range of actual expected pH change in gray.) Five parameters are included: calcification, metabolism, growth, fertility, and survival. Note that the overall trend is positive for all studies up to 0.30 units of pH reduction.

CHAPTER 11

Mass Walrus Death from CO_2 – Another Fake Catastrophe from Sir David Attenborough

This will be a very short chapter as it doesn't take long to expose another fraud perpetrated by Sir David Attenborough. In this case it's the hundreds of walruses falling to their deaths from cliffs, allegedly due to climate change (see Fig. 94).

In early April of 2019 Netflix premiered an episode of its eight-part series *Our Planet* titled "Frozen Worlds." It featured a dramatic scene of walruses falling 76 meters (250 feet) to their death onto the rocks and onto the other walruses below. The series was hosted by Sir David Attenborough who made it quite clear that this was caused by climate change. In the documentary he states, "Their natural home is out on the sea ice" and "they do so (haul out onto the beach) out of desperation, not out of choice."

This is an outright lie. The natural home of walruses is in the sea, just like it is for their pinniped cousins, seals and sea lions. These species haul out variously on ice or land to give birth and to rest. But their hunting grounds are in the ocean. They do not haul out on land "out of desperation," they do it out of choice for very good reasons and have been doing so since their ancestors evolved more than 20 million years ago.[1]

---

1. "Pinniped," Wikipedia, October 5, 2020. https://en.wikipedia.org/wiki/Pinniped#Evolutionary_history.

Figure 94. A screenshot from the "Frozen Worlds" episode of *Our Planet* of one of the hundreds of walruses that fell to their death. The entire episode is posted on YouTube. The walrus segment begins at minute 44:29.*

* Frozen Worlds," *Our Planet*, Netflix – YouTube. https://www.google.com/search?q=youtube+oor+planet+Frozen+Worlds.

While there is no doubt that polar bears require winter and spring ice to hunt seals, there is no hard evidence that walruses require ice at all. It is very likely that they haul out on ice when it is convenient, and especially when they are not faced with a pack of hungry polar bears. But unlike polar bears they do not need ice for hunting.

Attenborough claimed that global warming had melted the ice the walruses would normally have hauled out on and that now they must resort to hauling out on the shoreline where, due to lack of space, many had climbed a steep cliff to find a spot to rest. He states, "They struggle up the 80-meter cliffs."

Walruses cannot climb up cliffs (see Fig. 95, page 200). Then Attenborough claims: "A walruses' eyesight is poor out of the water." So, somehow the walruses had sufficient eyesight to waddle the long distance to get to the top of the cliff, but then they just fell off the cliff edge by the hundreds because they couldn't see it?

The location of this incident was not reported in the film. Fortunately, there were a few people who knew about this occurrence and had an entirely different tale to tell. Dr. Susan Crockford of polarbearscience.com recalled the reports from Ryrkaypiy on the Chukotka

Figure 95. The colony of about 5,000 female walruses and their cubs hauled out in their refuge near Ryrkaypiy on the Chukotka coast of northern Russia. There were so many of them some had made their way up the slope on the left, finding themselves poised above the cliff. It seems the polar bears took advantage of this by rushing them, so they had no other escape. It is not uncommon for polar bears to hunt walruses (see Fig. 96).

Figure 96. A polar bear attacking a walrus in hopes of making a meal or two. It is much easier to frighten the walruses to jump off a cliff to their death and then eat the carcasses, but that opportunity does not often present itself.

coast of northern Russia in October of 2017.[2] A colony of about 5,000 female walruses and their cubs regularly hauled out at Kozhevnikova Cape not far from the town of Rurkaypiy. The *Siberian Times* reported on October 17, 2017 with a headline that read: "Village besieged by polar bears as hundreds of terrorized walruses fall 38 meters to their death" (it appears that Attenborough exaggerated the height of the cliff by more than double). The story began with:

> *Around 20 beasts have surrounded Ryrkaypiy, with one bear cub trying to get into a house through the window. The polar bears were attracted by 5,000 walruses that appeared this year at a special protection zone in Chukotka. Many of the frightened flippered marine mammals fell off cliffs at Kozhevnikova Cape as they sought to flee the invaders. Several hundred fell to their deaths, and the polar bears then ate the carcasses.*
>
> *Head of WWF* (World Wide Fund for Nature) *project Polar Bear Patrol, Viktor Nikiforov, said: 'This autumn the situation is alarming. Many crashed, falling from a height. Their rookery had attracted polar bears. The walruses were obviously frightened by the predators, panicked and fell from the top to their deaths. Many crashed, falling from a height. Their rookery had attracted polar bears.'*[3]

That seems pretty straightforward. And the photograph of the walrus colony showing the cliff where hundreds fell to their death tells the story. The walruses did not "struggle up the 80-meter cliff." They waddled up a relatively gentle slope and thus arrived at the top of the cliff (see Fig. 95).

But even if the polar bear explanation were somehow incorrect, which seems unlikely, it is not the only such incident known. The US Fish and Wildlife Service reported a very similar situation that occurred for three years in a row at a walrus colony of more than 12,000 on the beach at Cape Pierce in the Togiak National Wildlife Refuge in Alaska. Beginning in 1994, and for two more years, numerous walruses fell down a cliff to their death. The people who witnessed it were unable

---

2. Susan Crockford, "Attenborough's tragedy porn of walruses plunging to their deaths because of climate change is contrived nonsense," April 7, 2019. https://polarbearscience.com/2019/04/07/attenboroughs-tragedy-porn-of-walruses-plunging-to-their-deaths-because-of-climate-change-is-contrived-nonsense/.
3. *Siberian Times* reporter, "Village besieged by polar bears as hundreds of terrorized walruses fall 38 meters to their death," the *Siberian Times*, October 19, 2017. https://siberiantimes.com/ecology/others/news/village-besieged-by-polar-bears-as-hundreds-of-terrorised-walruses-fall-38-metres-to-their-deaths/.

to determine the cause.[4] One probable cause is a rapidly growing population. As with polar bears there are strict regulations regarding the hunting of walruses, and they are effectively restricted to aboriginal hunters who live in the Arctic. In addition, many colonies are completely protected, including the colony at Kozhevnikova Cape where the walruses fell to their death.

It seems quite clear that polar bears, not climate change, caused the walruses to choose the cliff rather than to be attacked by the 20 bears (later reports claim there were 38 bears in the group). But that is not the biggest deception in the fraudulent and sensationalized reporting of Netflix's *Our Planet*. The reason walruses have those huge tusks is so they can dig for clams and other species such as worms, gastropods, cephalopods, crustaceans, and sea cucumbers. In other words, they are bottom feeders, similar to Sir David Attenborough and the film-crew of *Our Planet*.

Walruses can only dive to a depth of about 91 meters (300 feet) and more typically dive to depths only half of that. Therefore, when the ice recedes northward towards the pole from the northern coast of Russia in the summer and early fall, there is no ice in the shallow ocean near the coast. At the onshore area where they hauled out the ocean is shallow enough for the walruses to dive to the ocean floor to feed. And that is precisely why they hauled out from there, as they have done for many years. That's probably why it is designated as a walrus sanctuary.

Walruses are not an open-ocean species but rather a coastal species, and when the ice recedes northward in the summer they remain on the coast where the fishing is good. It is a real shame that Sir David Attenborough takes advantage of the average person's lack of knowledge about obscure details concerning nature, which he is fully aware of.

Here is what the World Wide Fund for Nature (WWF) states about walrus habitats:

> *Walruses are widely distributed but occupy a relatively narrow ecological niche, requiring areas of shallow water with bottom substrates that support a productive bivalve community, the reliable presence of open water to access these feeding areas, and suitable ice or land for hauling out.*[5]

---

4. Grant Burns, "Walruses Falling off Alaska Cliffs" undated video with interviews. https://www.dailymotion.com/video/x2m72ze.
5. J. W. Higdon and D. B. Stewart, "State of Circumpolar Walrus Populations," WWF, May 2018. https://arcticwwf.org/site/assets/files/1541/walrus-report-2018-screen.pdf.

Figure 97. Here you can see the line of walruses offshore on their feeding grounds. Walruses are very restricted in their habitat as they can dive no more than 90 meters deep (300 feet) but most of their feeding is done between 10 to 50 meters (33 to 165 feet) to feed on clams and other bottom species. That is why they haul out onto shore as the ice recedes northward in the summer.

This validates the conclusion that walruses will use either ice or land to haul out on so long as there is sufficient open water and it is shallow enough for them to dive to the bottom to forage for food (see Fig. 97).

As usual, Dr. Susan Crockford has the most credible analysis of the history of walrus haulouts on the beaches of Alaska and the Russian Arctic:

> *Recent mass haulouts of walrus females and calves on the beaches of Alaska and Russia bordering the Chukchi Sea have been blamed by US government biologists and WWF activists on lack of summer sea ice, claims that have been amplified into alarming scare stories by a compliant media and embellished with alarming stories of trampling deaths. However, such claims ignore the published literature documenting previous events, which suggest a different cause. Rather than lack of ice, the presence of such massive herds onshore in six out of the eight years since 2007 indicate that the now well-protected walrus population may be so high that it is approaching the carrying capacity of its habitat. Sea ice maps for the months when known mass haulouts occurred, compared to years when they did not, suggest no strong correlation with low sea ice levels. Instead of there being a clear case for blaming this walrus behavior squarely on global warming, the evidence suggests that high population numbers may be a significant factor, among other potential triggers not fully understood. Those who suggest these events are a sign of pending*

catastrophe are looking for victims of global warming to tally on a ledger, but in doing so they not only fail to acknowledge potential consequences of natural fluctuations in walrus population size but fail to concede the obvious resilience of this species to profound sea ice changes they have survived repeatedly before now.[6]

Walruses, like polar bears, are so remote from most human populations that they might as well be invisible to most people. Unfortunately, there are a lot of activists, media outlets, politicians, and scientists that take advantage of this fact and use it to their own benefit. Let us hope that this powerful convergence of interests, that is aimed at building fear and guilt in adults and children alike, is not ultimately successful.

In the final analysis it is the duty of highly regarded persons, such as Sir David Attenborough, to stick to the truth and not to sell their soul by making apocalyptic predictions that they know are based on falsehoods. I personally challenge Sir Attenborough to dispute the points presented herein regarding seabirds, plastic, walruses, and polar bears. I look forward to his rebuttal.

And if you want a definitive counter response to Attenborough's propaganda take a look at the Global Warming Policy Forum's video that eviscerates his narrative to such an extent that you must reject claims of a "climate change catastrophe" having anything to do with either walruses or polar bears.[7]

---

6. Susan Crockford, "On the Beach – Walrus Haulouts are Nothing New," The Global Warming Policy Foundation, 2014. https://www.thegwpf.org/content/uploads/2014/10/walrus-fuss.pdf.
7. Susan Crockford, "Netflix, Attenborough and cliff-falling walruses: the making of a false climate icon," GWPF, May 17, 2019. https://www.youtube.com/watch?v=IatVKZZcPGO&list=LLzWji6QiQmCEoU6vIigmceA&index=201.

Epilogue

The COVID-19 pandemic has clearly demonstrated that there are real catastrophes caused by microscopic viral agents, and that even this has been somewhat overplayed in an overly risk-averse society. Hopefully this will help bring the fake catastrophes into perspective. People are not dying by the tens of thousands from climate change. Species are not going extinct by the tens of thousands either. And genetically modified food has not been known to cause a single illness, never mind thousands of deaths.

Lack of beta-carotene is still causing hundreds of thousands of deaths and cases of blindness every year, mainly in children. The Philippine's approval of Golden Rice for food and feed is hopefully the beginning of its adoption in all the countries that need it. But this is not yet certain as Greenpeace and their ilk continue to campaign against it.

Meanwhile the whole Earth, except for the ice-covered parts which are thankfully shrinking for now, is greening on a daily basis. Life expectancies continue to rise in most countries due to a combination of better diet and health care, as well as a supply of reliable and cost-effective energy from fossil fuels, hydroelectric, nuclear, and biomass. If common sense prevails wind and solar energy will eventually be phased out except for off-grid applications.

I will leave readers with one last graph, the global consumption of all energy sources for all purposes (see Fig. 98). Increasing global nuclear energy by about 40 percent would more than replace unreliable, heavily subsidized wind and solar with clean reliable energy. This would also displace the fossil fuels required to back up wind and solar most of the time.

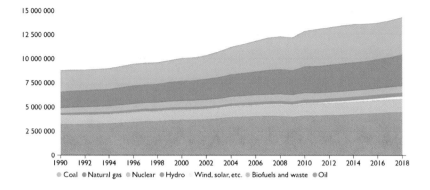

Figure 98. Total global energy supply from all sources expressed in thousands of tons of oil equivalent. Note that biomass and waste-to-energy produce 4.6 times as much energy as wind, solar, etc. If all unwanted combustible waste globally were converted to reliable energy it may well be able to displace unreliable wind and solar and greatly reduce the need for landfills. This graph is interactive on the International Energy Agency website.[1]

---

1. "Data and Statistics," International Energy Agency, 2020. https://www.iea.org/data-and-statistics?country=WORLD&fuel=Energy%20supply&indicator=Total%20energy%20supply%20(TES)%20by%20source.

EPILOGUE　　　　　　　　　　　　　　　　　　　　　　　　　207

Made in the USA
Monee, IL
01 April 2021